T5-CVE-061

Illustration on Cover and Title Page:

Dajazmach Ganama Delnessaw, a lesser early twentieth century Ethiopian chief, with his family and followers.

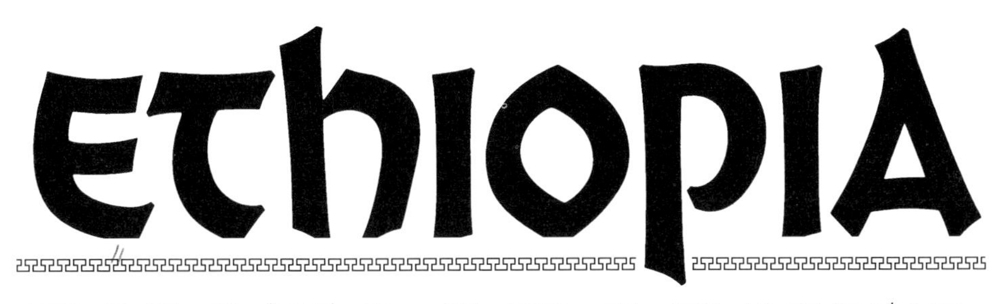

PHOTOGRAPHED

Historic Photographs of the Country and its People taken between 1867 and 1935

Richard Pankhurst
&
Denis Gérard

Kegan Paul International
London and New York

First published in 1996 by
Kegan Paul International
UK: P.O. Box 256, London WC1B 3SW, England
Tel: (0171) 580 5511 Fax: (0171) 436 0899
E-mail: books@keganpau.demon.co.uk
Internet: http://www.demon.co.uk/keganpaul/
USA: 562 West 113th Street, New York, NY, 10025, USA
Tel: (212) 666 1000 Fax: (212) 316 3100

Distributed by
John Wiley & Sons Ltd
Southern Cross Trading Estate
1 Oldlands Way, Bognor Regis
West Sussex, PO22 9SA, England
Tel: (01243) 779 777 Fax: (01243) 820 250

Columbia University Press
562 West 113th Street
New York, NY 10025. USA
Tel: (212) 666 1000 Fax: (212) 316 3100

Set in Korinna Medium
Printed in Great Britain by T.J. Press, Padstow, Cornwall

British Library Cataloguing in Publication Data

Pankhurst, Richard, 1927 –
Ethiopia photographed: historic photographs of the country and its people
taken between 1867 and 1935
1. Ethiopians – Portraits 2. Ethiopians – Pictorial works
3. Ethiopia – Social life and customs – Pictorial works
I. Title II. Gérard, Denis
963'.04'0222
ISBN 0-7103-0504-4

Library of Congress Cataloging-in-Publication Data
Applied for

Contents

ACKNOWLEDGEMENTS

THE EDITORS WISH TO THANK the many individuals and institutions that made this volume possible. Those who kindly lent photographs for reproduction include H.E. Ato Tekle Tsadik Mekuria, Ato Gabre Mariam Tessema, Woizero Susie Workneh, Woizero Eleni Mukria, Dejazmach Fikre Selassie, Ato Dedria Mohomed, Ato Salah ed-Din Mohamed, Ato Hashim Ali, Dr Yohannes Kenfu, Woizero Alem, Tsehai Iyasu, Bertrard Hirsch, Michel Perret, Woizero Rosa Terefe, Ato Mangasha Geneme, Woizero Kiddest Wossenyalesh, Ato Habib Alimira, Ato Hassen Yayo, Ato Tadele Yidnakachew, Ato Boyalu Getachew, Ato Belai Gidey, M. Michel Pasteau, M. Alain Le Seac'h of the French Ministry of Foreign Affairs, Ambassador James Glaze, Mrs Merrill Christopher, Mrs Patricia Fenn, Mr Ian Campbell, Mr Mohammed Moubarek of the Indian community, Mrs Vergine Barssiguian Karibian, Russell Harris, MS Chris Wright of the Royal Anthropological Institute in London, Ato Seyoum Abate, of the Addis Ababa Museum, and last but not least, Ato Demeke Berhane, of the Institute of Ethiopian Studies. The Institute today houses the most comprehensive collection of Ethiopian photographs in the world, and grateful thanks for permission to draw on its holdings are due to its successive directors, Dr Tadesse Beyene, Professor Bahru Zewde and Dr Abdussamad H. Ahmad. Invaluable help in identifying photographs was received from Lij Michael Imru, Dr Berhanou Abebe, Ato Ahmed Zakaria, Ato Fassil Ghiorgis, Mr Avedis Terzian, and Mr Sahag Boghossian.

Valuable information on photographic holdings in various parts of the world were kindly provided by numerous friends and colleagues, most notably Mrs Chris Rosenfeld, of Washington, Dr Bairu Tafla, of Hamburg, Signor Alberto Imperiali, of Rome, and M. Patrick Groult, of Paris. Information on foreign collections of Ethiopian photographic collections was also contributed by Dr Elizabeth Edwards, of the Pitt Rivers Museum in Oxford, Dr Silvana Palma of the Istituto Italo-Africano, Rome, M. Philippe Revol of the Musée de l' Homme and M. Michel Florin of the Société, both of Paris, Mr Russell Harris of the Victoria and Albert Museum's Layfayette project and Joanna Scudden of the Royal Geographical Society, both of London, and Derek Howlett of the Powell-Cotton Museum, of Birchington, Kent.

An Ethiopian warrior.

INTRODUCTION

IMAGES OF THE COUNTRY AND ITS HISTORY

HISTORIC ETHIOPIA, situated in lofty, often inaccessible, mountains between the Red Sea and the Blue Nile, and extending far into the Horn of Africa, was an ancient country, internationally long renowned. Since time immemorial it exercised an extraordinary fascination on the outside world. Ethiopia was also a complex, and at times mysterious country, which many foreigners found, and indeed still find, impossible to understand. The country appeared moreover to different peoples - and at differents times - in different, and frequently changing, guises. At the same time it was - and remains to this day - a highly photogenic country, as we hope the following pages will reveal.

Before looking at the photographs, and at the history of the photographers who took them, we attempt to introduce the reader to a few highlights of Ethiopia's long history, and see how foreign images of the country, as in a kaleidoscope, underwent many a great transformation.

Ethiopia owed its importance in ancient times to the fact that it was the land where the mighty Nile had its source, and lay beside one of the world's most important international trade routes - the route through the Red Sea and Gulf of Aden, a route which linked the Mediterranean, and hence the West, with India, China and other countries of the East.

The ancient Egyptians, who provide our first important testimony on the region, knew the area on both counts. The Ethiopian highlands, they realised, were the source of the Nile, to whose water - and silt, Egypt owed its very existence. At the same time the low and arid country towards the coast constituted the core of the Land of Punt, whence the Pharaohs obtained myrrh and other gums and resins necessary for their devotions, on which account they spoke of the country as God's Land. The region was so important that many Pharaohs despatched naval expeditions to its coast. One of the most famous was sent around 1495 B.C. by Queen Hapshetsut. She records in hieroglyphs on the walls of her great temple at Thebes in Upper Egypt that she had been commanded by the god Amon to obtain and plant myrrh trees, and thereby "establish a Punt for him" in her own country. The expedition, as depicted in her temple, returned home with the produce of the area: ivory, gold, incense, and animal skins, as well as some local people, perhaps as slaves.

Trade with the Ethiopian and Horn of African region seems to have continued into Biblical times. This is suggested by the fact that Ophir, whence King Solomon is said in the Book of Kings to have sent an expedition in quest of gold, incense and precious stones, was very probably in the area.

Ethiopia and the Ethiopians also figure in the Bible, which contains no less than thirty-three references to the country or its inhabitants. In one poetic passage in the Book of Zephania, God speaks of lands "beyond the rivers of Ethiopia", while Psalm 68 contains the long to be emotive prophecy: "Princes shall come out of Egypt; Ethiopia shall soon stretch out her hands unto God".

The Ptolemies, who ruled Egypt in the half millennium B.C., were likewise deeply interested in the southern Red Sea coast of Africa, to which they, like the Pharoahs before them, despatched numerous expeditions. These were sent largely in quest of elephants, which were needed for war, and have aptly been called the "tanks of the ancient world".

The ancient Greeks were also well aware of the region. Long before the Christian era they invented the name Ethiopia, and used it to refer to the country inhabited by "burnt", or sun-tanned, people south of Egypt. Homer in the ninth century B.C. spoke of the people of Ethiopia as "the most distant of men", living "at the earth's two verges, in sunset lands, and lands of the rising sun". He believed, like many of his compatriots, that the Ethiopians were the best people in the world. He referred to them as the "blameless Ethiopians", and claimed that their country was visited by the Gods, who "lingered delightedly" there.

Meanwhile in North-East Africa itself the half millennium or so before the birth of Christ witnessed the rise of an important, and highly developed civilisation in what is now northern Ethiopia. The inhabitants of this area accepted the same pantheon of gods as were worshipped in South Arabia, including the sun, the moon, and the goddess Venus. The first capital in what is now Ethiopia was, as far as we know, at a place called Yéha, which was in its day a settlement of considerable size and significance.

This is evident from the city's huge stone temple, measuring twenty metres long by fifteen wide, whose ten metre high walls can be seen - and photographed - to this day.

The centre of political power subsequently shifted a few days' journey westwards to nearby Aksum, which developed as an even greater city, and the capital of an extensive kingdom, destined to become the most important state between the Roman Empire and Persia.

The Aksumite realm, which had access to the sea at the Red Sea port of Adulis, was a major commercial kingdom. It traded with lands as far afield as Arabia, Egypt, Persia, India, and even Ceylon. Aksum for several centuries also produced its own currency in gold, silver and bronze, with inscriptions in either the local Aksumite language, Ge'ez, or else in Greek, the then international language of the Red Sea area. Most of these coins bore effigies of the ruling monarch, and sometimes representations of a sprig of grain, and popular slogans, such as "May it Please the People".

The first glimpse of Aksumite foreign trade is provided by the *Periplus of the Erythraean Sea*, an early Graeco-Egyptian commercial text written in the first century or so of the Christian era. This work shows that Aksum's exports, via the Red Sea port of Adulis, included ivory and rhinoceros-horn, as well as tortoise-shell from the coastal waters, while the Roman author Pliny at about the same time also mentioned the export of slaves. Imports, according to the *Periplus*, consisted of large quantities of Indian cloth, besides iron and brass, plates, knives and cups, and sundry other manufactured and luxury goods, mainly from India, Egypt and the Mediterranean area.

The Aksumites were remarkably skilled artisans. They built magnificent palaces, temples and churches, a large dam, several great funeral vaults, and numerous obelisks, in some instances beautifully carved. One of the latter, no less than thirty-three metres long, has the distinction of being the longest block of stone ever excavated by man. Over the centuries these great stelae have been regarded with admiration by countless visitors, many of whom took excellent photographs of them. One, looted in 1937 on the personal orders of the fascist dictator Benito Mussolini, was taken to Rome, and, despite Ethiopian requests for its restitution, has thus far not been returned.

Because of its external relations the Aksumite kingdom came into contact with Christianity at a very early period. Ethiopian church tradition holds that some Ethiopians were already converted at the time of the Apostles. Be that as it may, there is irrefutable evidence that Christianity had become the state religion of the kingdom by around A.D. 300. The country's main conversion took place, according to early Byzantine evidence, when a ship carrying a group of Syrian Christians returning from India was boarded off the Red Sea coast on account of an earlier dispute. Virtually all the voyagers were put to the sword, but two young boys survived, and were found studying, doubtless the Bible, under a tree. They were brought to the Aksumite king, who appointed one of them his cup bearer, and the other, by name Frumentius, as his secretary and treasurer.

The king, we are told, died shortly afterwards, leaving his wife with an infant as the heir to the throne. She begged the two young men with tears in her eyes to stay with her to help her govern the kingdom, at least until her son should come of age. They duly agreed. Frumentius, who later reportedly held the reins of government, sought out visiting Christian merchants, and together with them established Christian churches, and did much to expand the new religion.

Frumentius subsequently travelled to Alexandria, then the principal centre of Christianity in the East, where he informed the great Patriarch Athanasius what had thus far been done for the faith at Aksum, and begged him to despatch a Bishop. Athanasius, having heard what Frumentius had already done, replied, "What man can we find other than you, who can accomplish these things?" He thereupon consecrated the young man as Ethiopia's first bishop. Frumentius then returned to Aksum, where he adopted the name of Abba Salama, or Father of Peace.

The young monarch whom Frumentius had helped govern was apparently King Ezana. He was, as far as we know, the first Aksumite ruler openly to avow the Christian faith. This is evident from the fact that some of his coins, thought to be the earlier ones, bore representations of the old South Arabian deities, the sun and moon, while others, doubtless struck after his conversion, bear the Cross of Christ. They are of special interest in that they were the first coins in the world to bear that symbol. The coming of Christianity is also apparent in Ezana's inscriptions, the earliest of which referred to pagan gods while a later one paid homage to the presumably Christian Lord of the Universe.

At this time, and in the next few centuries, the Bible, and other religious texts, were translated into Ge'ez.[1] The development of Christianity also owed much to the coming, in the late fifth and early six centuries, of the so-called Nine Saints, holy men from Syria and its environs, who founded a number of important monasteries in what is now northern Ethiopia. Church schools, which were to remain the basis of the country's education for a millennium and a half, were also established around this time.

This period also witnessed extensive church building, which included a particularly characteristic Ethiopian phenomenon: the excavation in Tegray and elsewhere of underground or semi-underground churches cut out of the living rock.

Ezana and the kings who followed him undertook a number of major military expeditions, which greatly expanded the frontiers of the Aksumite state, particularly southward and westward. Ezana, a great warrior king, on one occasion marched far into what is now Sudan up to the Nile. He recorded his victories in a number of stone inscriptions, which can be viewed in Aksum to this day. Some are written in Sabaean, the old language of South Arabia, others in Ge'ez or Greek. King Kaléb, one of his successors, later despatched a no less notable sixth century expedition across the Red Sea into South Arabia.

One of many foreign merchants to visit both Aksum and the port of Adulis in this period was a Greek-speaking Egyptian, Kosmas Indicopleustes, who later became a monk. A man of scholarly disposition he wrote an important work, the *Christian Topography*, which was, and remains, of unique interest. It gave the outside world its first eye-witness account of Ethiopia, which was then, its author insists, an integral part of the Christian world, as well as a major commercial emporium, trading with countries as far away as Egypt, Persia, India and Ceylon.

The Aksumite kingdom was particularly well known to the Arabs on the opposite side of the Red Sea, who referred to it, and to Ethiopia in general, as Habash. The term was derived from the name of a coastal tribe, the Habashat, and was in turn the root of the English word "Abyssinia", the French "Abyssinie", the German "Abessinen", and similar names in other European languages. Since the Arabic word *habash* also means "mixed" Ethiopians tended to consider the name derogatory, and by and large never spoke of their country as Abyssinia, but always as Ethiopia.

The Prophet Muhammad was one of many Arabs familiar with Habash. He had been brought up by his grandfather, Abdal Muttalib, who had travelled to the country on business, and his nurse was reportedly an Ethiopian, who taught him some words of Ge'ez. Later, when his early followers were being persecuted in Arabia, the Prophet, according to Muslim tradition, pointed towards Abyssinia, and, echoing the ancient Greeks, declared, "There, is a land where no one is wronged, a land of righteousness. Go there", he added, "and remain there until it pleases the Lord to open the way". A small band of Muslims duly made their way to the Aksumite kingdom, where their enemies despatched an embassy with costly gifts to demand their extradition. The then Aksumite ruler, King Armah, however, interrogated the exiles, and apparently finding no harm from his point of view in their ideas, turned to the emissaries from Arabia, and replied in a justly famous riposte, "Even if you would offer me a mountain of gold I would not give up these people who have sought refuge with me".

The Prophet, according to Muslim tradition, later requested Armah to betroth him to one of the woman refugees, Umm Habibah. The monarch did so, and gave her a golden dowry. From her and another of his wives, Umm Salma, who had also found asylum in Ethiopia, Muhammad is said to have learnt of the beauty of Aksum's principal church, St Mary of Seyon, which was dedicated to the Virgin Mary. Tradition also holds that the Prophet, when condemning the representation in painting of the human form, placed his hand over an icon of the Virgin, which in all probability had come from Aksum, and thus protected it from the ban. After Armah's death he reportedly prayed for the Aksumite king's soul, and commanded his followers to "leave the Abyssinians in peace", thereby exempting them from the Jihad, or Holy War.

Ethiopia's close relationship with early Islam also found expression in Muhammad's choice of an Ethiopian, by name Bilal, as his first Muezzin, who called the faithful to prayer. The Prophet spoke of him endearingly as "the first fruit of Abyssinia". Contact between Ethiopia and Islam is likewise graphically reflected in the Qoran which contains a number of Ge'ez loan-words.

Abyssinia in later centuries was well known to the great Arab geographers of the Middle Ages. Several referred to the country in their writings, and provided the outside world with valuable glimpses of the country, and more especially of the up to then little known Muslim sultanates, among them Ifat and Adal, located to the east of the Christian empire.

This period, and the centuries which followed, coincided with major changes in Ethiopia. The Aksum kingdom began to decline around the middle of the seventh century, and coins ceased to be minted. Later, around the tenth century, the centre of political power shifted southwards to the mountains of Lasta, where a new ruling dynasty, the Zagwé, emerged as the inheritors of the Aksumite Christian kingdom. Situated however much further from the sea the Zagwé state had far fewer contacts than the Aksumites with the outside world.

Perhaps the greatest of the Zagwé rulers was King Lalibela, who, reputedly wishing to make his land-locked capital, Roha, replace Jerusalem as a place of pilgrimage, built eleven remarkably fine rock-hewn churches. Several of them deliberately embodied Aksumite motifs, and one of them, Madhané Alam, was perhaps designed as a replica of the old church of St Mary of Seyon. After his death, Roha was renamed Lalibela in his honour. His churches, which have been ranked among the wonders of the world, remain to this day a great, and very photogenic, tourist attraction.

Power in Ethiopia later once more shifted south, when around 1270 the Zagwé were overthrown, and replaced by a new dynasty based in Shawa, the province in which Addis Ababa is now situated. The new ruling family claimed descent from the Biblical King Solomon and his son, by the legendary Ethiopian Queen of Sheba.

This Solomonic period, as it has been called, witnessed a great new flourishing of Ethiopian Christian civilisation. Several major works of Ge'ez literature were written at this time. They included the first of the royal chronicles, which described the reigns of successive Ethiopian emperors, and the country's traditional code of law, the *Fetha Nagast*, or Law of the Kings, which was translated from an Arabic text by an Egyptian Copt, and proclaimed the Divine Right of Kings.

No less important was the *Kebra Nagast*, or Glory of Kings, written by a priest from Aksum, which became the country's national epic. This work, which did much to reinforce royal authority as specified in the *Fetha Nagast*, told the story of the Queen of Sheba, an Ethiopian woman ruler called Makeda, who reputedly travelled to Jerusalem to learn of the wisdom of Solomon. The text relates that the Biblical monarch tried to seduce her, but that she refused his advances. He thereupon promised that he would not take her by force, provided that she for her part promised not to steal anything from his palace. To this she readily agreed. He then prepared the famous spicy banquet, and, after eating it, they both went to bed, separately. Not long afterwards the queen awoke, very thirsty, and drank some water. Solomon thereupon seized her, claiming that she had broken their agreement. She replied that she had taken only a sip of water, to which he answered that nothing in the world was more valuable than water, which she could not deny, after which he "wrought his will with her". Solomon, it is claimed, later had a dream in which he saw the sun, which had hitherto shone over Israel, move away, to shine instead over Ethiopia.

The queen, it is said, subsequently returned to Ethiopia, where she gave birth to a son, called Menilek, the reputed founder of the Ethiopian royal dynasty. The text goes on to state that this prince later travelled to Israel to see his father, that the latter recognised him as his "first-born" son, and wished him to remain in Jerusalem with him. Menilek, however, insisted on returning to his mother, whereupon Solomon gave him the first-born sons of the leaders of Israel to accompany him. The young men, reluctant to travel to a land far away from their Temple, are said to have then secretly purloined the Ark of the Covenant, to take with them to Ethiopia. When Menilek learnt of their success, he declared that it must have been God's wish that the Ark should be taken to Ethiopia - where tradition holds that it still exists.

The message embodied in this long story, as spelt out in the *Kebra Nagast*, was a powerful one: It was that God had transferred his love from Israel to Ethiopia, that the Ethiopians had replaced the Jews as His Chosen People, and, more specifically, that the rulers of Ethiopia were descended directly from the Biblical Kings of Israel - and hence entitled to all the authority, obedience and respect due to such an august God-given heritage. Subsequent Christian rulers all claimed descent from Solomon, and may therefore be described as members of a "Solomonic dynasty".

Ethiopian Christians, it may be noted, had all this time close ties with Judaism. Paying great honour to the Old Testament they followed, and indeed generally still follow, all the Mosaic dietary rules, practice circumcision, and for hundreds of years observed a Biblical Saturday Sabbath. Every Ethiopian Christian church likewise houses as its most cherished content a *tabot*, or altar slab, which symbolises the Mosaic Ark of the Covenant, which, many believe, is currently preserved in Aksum.

The Shawa-based Christian monarchy, unlike its Aksumite and Zagwé predecessors, had no fixed capital. The custom of the Solomonic rulers, as a chronicler later wrote, was to travel with their armies from place to place until their last resting place, that is to say where they were buried. The monarchs of this period travelled partly in quest of supplies, partly to govern the far-flung empire, and partly to prosecute their wars. Many of these over the centuries were fought against the lowland sultanates to the east, which had by then espoused Islam.

Though located far in the interior, the Ethiopian Christians of this period, like their ancestors, were not forgetful of the outside world. They regarded Christian Jerusalem with deep reverence, went there on pilgrimage if at all possible, and were well aware that they belonged to a wider Christendom, by this time situated largely in Europe. So far from regarding themselves as on the periphery of the Christian Universe they considered themselves as in a sense at its centre, but nevertheless yearned for communion with their co-religionaries in other lands.

This desire for contact with Christian Europe was intensified as a result of the introduction of firearms into the Red Sea and Gulf of Aden area in the fifteenth century. The country's rulers, who were relatively isolated from the coast where such weapons were more readily available, became thereafter increasingly conscious of their military deficiency, and realised that it could be remedied only by the import of cannons and rifles from Europe, and, if possible, soldiers to fire them or instruct Ethiopians in their use.

Medieval Europe's perception of Ethiopia owed much to the advent in Jerusalem of numerous Ethiopian Christian priests, monks and other pilgrims. Their importance was recognised in 1189 by Sultan ad-Din, the ruler of Egypt, better known in the West as Saladin, who, after driving the Latin clergy from the Holy City, granted the Ethiopians the Chapel of the Invention of the Cross in Jerusalem and a station in the Grotto of the Nativity in Bethlehem.

Information about the presence of Ethiopians in the Holy Land, and of the Christian country from which they came, gradually percolated to Europe. This gave rise to the hope that the then mysterious Ethiopian kingdom might prove a valuable ally in the Christian Crusades against Islam. One of the first Europeans to urge the desirability of such an alliance was a Dominican monk, Guillaume Adam. After several unsuccessful attempts to enter the country he drafted a plan in 1317 for the King of France in which he argued that Ethiopian co-operation would enable European Christendom to eradicate the hated Saracens.

The idea of a Grand Alliance against Islam was reinforced in the West by the legend of a powerful Christian ruler called the Prester John, which had been circulating since the thirteenth century. It was originally believed that this monarch lived in or near India, and that his primary aim was the liberation of the Holy Sepulchre from the infidels. Support for this belief was found in a fictitious letter, supposedly from this ruler, which attracted considerable interest, and was translated into several European languages. Numerous efforts were made to locate the non-existent ruler. When they failed, belief in a Prester John located in Asia gave way to the idea that he was in fact none other than the Christian emperor of Ethiopia, whose existence had by then become well known.

Contact between Ethiopia and Western Europe, and especially Italy, grew closer in the late fourteenth and early fifteenth centuries. A Florentine merchant, Antonio Bartoli, entered Ethiopia around the 1390s, and a Sicilian, Pietro Rombulo, in 1407. The latter spent almost four decades in the country before being sent by its ruler on an embassy to India and China. Rombulo subsequently returned to Italy as part of a mission from the notable Ethiopian Emperor Zar'a Ya'qob. The first European embassy to Ethiopia had meanwhile been despatched by a French nobleman, the Duc de Berry. Another Ethiopian monarch, Emperor Yeshaq, wrote in 1428 to King Alfonso of Aragon, proposing an alliance against Islam, to be cemented by a double dynastic marriage. This failed to materialise, but Alfonso later despatched a message to Emperor Zar'a Ya'qob, in 1450, promising to send him artisans whom the latter had apparently requested.

Europe's awareness of Ethiopia had meanwhile been heightened by news that an Ethiopian delegation was to attend the Ecclesiastical Council of Florence in 1441. The embassy, it later transpired, consisted of only two monks, who came in fact not from Ethiopia, but from the Ethiopian community in Jerusalem. Their arrival nevertheless created considerable interest in Italy, and was important in furthering knowledge of their distant motherland. European interest in Ethiopia was by then spreading. The Venetians, trade rivals of the Egyptians, wanted to develop relations with the country, while the Spaniards and Portuguese also saw advantages for themselves in contacts.

The extent of Europe's knowledge of Ethiopia around this time is apparent from the data on the country found on two important maps. Pietro del Massajo's *Egyptus Novelo*, produced in Florence around 1454, and Fra Mauro's more famous *Mappomondo*, designed in Venice in 1460. Both contained a considerable amount of information on the country, doubtless gleaned from Ethiopian ecclesiastics visiting Italy. Several of the latter subsequently attracted the attention of a Venetian writer, Allesandro Zorzi, who interviewed them, and produced valuable itineraries of the routes they had taken to and from their native land.

A steady flow of Ethiopians travelling around this time to Italy made their way to Rome, where they attached themselves to one of its churches, the Church of St Stephen, which came to be called Santo Stefano degli Abissini. The presence of these Ethiopians served to expand European interest in their far-off country. One of those influenced was a German printer, Johannes Potken of Cologne, who, having heard them say Mass, in their liturgical language Ge'ez, or Ethiopic, set up the first printing press, for that language, and published a Ge'ez Psalter as early as 1513. The importance of the city's Ethiopian community was recognised two generations later, in 1539, when the Holy See purchased a hostel for them. Situated just behind St Peter's it has justly been called the Cradle of Ethiopian Studies in Europe.

Contacts between Ethiopia and Europe, and above all Italy, led meanwhile to the establishment in Ethiopia of a small, but not unimportant, foreign community. It consisted mainly of Venetians, Neapolitans and Genoese, and included two Venetian artists, Nicolo Brancaleone and Hieronimo Bicini, both of whom exerted some influence on the evolution of Ethiopian Christian art.

The Egyptians, most of whom had long since been converted to Islam, throughout this time also took a close interest in Ethiopia. This was not only, as in the past, because it was the country from which the Nile flowed, but also because it had become customary for the Ethiopians to import their Abun, or Patriarch, from among the Coptic clergy of Egypt. Ethiopia was the only country to which such Egyptian prelates were so despatched. Despite this link the two churches had differing beliefs and organisation. It is therefore entirely incorrect, as is sometimes done, to speak of the Ethiopian church as "Coptic".

The position of the Coptic prelates in Ethiopia was indeed a curious one. After their appointment to what was to all intents and purposes an exile post, they were treated in Ethiopia with every honour. However, they found themselves in an alien, even though Christian, land, with whose language, history and customs they were entirely unfamiliar. If they lived sufficiently long some of them overcame such difficulties, and gained considerable influence in their country of adoption. On at least one well documented occasion one of them even played a major role in selecting the Ethiopian heir to the throne.

The flow of the Nile from Ethiopia to Egypt, and the import of the Abun from Egypt to Ethiopia, meant that both countries were dependent on each other: Egypt for its material, and Ethiopia for its spiritual, life-blood.

The early sixteenth century was a major turning point in Ethiopian history. The Christian empire faced two interrelated threats. One was from the Ottoman Turks, who were expanding throughout the Red Sea area. The other was from the Muslim sultanate of Adal to the east, whose rulers, because of their proximity to the coast, had far easier access to firearms than the emperors of interior, and were therefore making successful annual armed incursions into the highlands.

Emperor Lebna Dengel, the Christian monarch at this critical stage, was a young and apparently far from wise ruler. Government was therefore at first in the hands of his redoubtable great-grandmother Empress Eléni. The daughter of the Muslim ruler of Hadeya, south-west of Shawa, she had become a Christian on marrying into the imperial family. Notwithstanding her Muslim origins she was alarmed by the threat of Turkish expansion, and accordingly instructed Matthew, a Christian Armenian merchant in her service, to travel to Portugal in 1509 with a letter to its ruler, King Manoel I, proposing an alliance against the Turks.

Matthew set forth as ordered, but encountered many difficulties, in part because his credentials were not accepted by the Portuguese, who could not believe that the Ethiopians, who were known to be dark-skinned, would have a white-skinned envoy. King Manoel, however, eventually decided on sending Matthew back to Ethiopia as guide to a Portuguese diplomatic mission. It reached the Red Sea port of Massawa in 1520, eleven years after Empress Eléni's original request for aid. To the Portuguese surprise Matthew on landing received a rapturous welcome from his Ethiopian co-religionaries and friends. Lebna Dengel, a proud young man, had, however, by then come of age. Confident that he could contain all his enemies, he was uninterested in a Portuguese alliance, and kept the mission in his country for no less than six years without concluding any agreement.

The protracted stay of the Portuguese from the point of view of European consciousness of Ethiopia was, however, of unique importance. The mission's chaplain, Francisco Alvares, wrote an extensive volume of memoirs, *Verdadera Informaçam das terras do Preste Joam das Indias*, or "Truthful Information about the Countries of the Prester John of the Indies". It constituted the first detailed account of Ethiopia, and dispelled much of the mystery which had hitherto enveloped the country in European eyes.

The ensuing period witnessed extensive fighting. The Muslim ruler of Adal, Imam Ahmad ibn Ibrahim, later nicknamed Gragn, or the Left-handed, launched a series of campaigns into the Ethiopian highlands in 1527. He succeeded in defeating Lebna Dengel, and overran the greater part of the empire. Innumerable Christians, as well as many followers of local traditional religions, particularly in the south-west, embraced Islam. The victorious Imam seemed to have destroyed the millennial-old Christian state. His triumph was described with a wealth of detail by his chronicler Shihab Ad-Din, in a memorable Arabic work, the *Futuh al-Habasha*, or Conquest of Abyssinia.

Ahmad's supremacy was, however, short-lived. Lebna Dengel succeeded in smuggling out an appeal for help to the Portuguese, who despatched a small, but powerfully armed, relief force. Led by Chistovão da Gama, son of the famous mariner Vasco da Gama, it landed at the port of Massawa in 1541, and contributed very materially to the Imam's defeat, and death in battle, in 1543. One of the Portuguese soldiers, Miguel de Castanhoso, later wrote a vivid account of the operation, and of the country in which he and his comrades had fought, and hoped to dominate.

The bitter conflict between the Christian empire and the Muslim sultanate, which had weakened both parties, meanwhile opened the way for yet another important event: the northward migration of the Oromo people, then better known as Gallas, who rapidly swept into the southern, western and eastern provinces. Their arrival was initially accompanied by much fighting, though many Oromos later rose to eminence in the Ethiopian Christian state. The Oromos for their part assimilated many of the southern peoples they encountered, though others fled northwards before them. The advent of the Oromos, and their subsequent incorporation into the Ethiopian empire, very significantly increased the country's ethnic, cultural and linguistic diversity, which is today much in evidence.

The Oromo advance was also important in that it contributed to another major shift in the centre of Ethiopian political power. The imperial capital was at this time moved from Shawa to the north-west of the realm, where a succession of short-lived capitals came into existence in the Lake Tana area.

The early sixteenth century meanwhile coincided with the rise of the great walled city of Harar in what is now south-eastern Ethiopia. The settlement, today a very photogenic place, expanded significantly after 1520 when a local Amir, Abu Bekr Muhammad, moved his capital there from Dakar, site of a nearby much older settlement. His rule was, however, soon cut short, for he was murdered five years later by the above-mentioned Imam Ahmad Ibrahim, or Ahmed Gragn, who in the course of his campaigns in the highlands brought considerable wealth to the city. He was succeeded by Amir Nur ibn al-Wazir Mujahid, who, it is said, erected the stout encircling walls which ever since then have been one the city's most dominant features.

Harar at that time and for the next three centuries was an independent and often militantly theocratic city state. Closed to Western eyes, but trading extensively with the East, particularly Arabia, Egypt and India, it issued its own currency for several hundred years. The city's many inhabitants included merchants, who travelled far and wide, as well as farmers who grew excellent coffee, and the mild stimulant chat (*Catha edulis*). Harar was also renowned for its handicrafts, which included weaving, basket-making and book-binding. Richard Burton, the translator of the *Thousand and One Nights*, who visited the town in the early nineteenth century, declared that its book-binding was surpassed only by that of Persia. Harar throughout its long history was also much renowned for its Islamic learning and scholarship.

As a result of their involvement in Ethiopian affairs some of the Portuguese, and their Spanish neighbours, conceived the idea of converting the country to Catholicism. Ignatius Loyola, founder of the Society of Jesus, was deeply interested in the country, and proposed to go there in person. He never did so, but his disciples, the Jesuits, were subsequently entrusted by the Pope with the task of conversion. A Jesuit mission led by André de Oviedo landed at Massawa in 1557, five days before its occupation by the Turks, who thereafter constituted a serious obstacle to any such foreign contacts. Notwithstanding the Turkish presence at the coast another, much abler, Jesuit missionary, Pero Paes, entered Ethiopia in 1603. By his zeal, and attractive offers of military support from Catholic Europe, he succeeded in winning the favour, and converting, two successive emperors, Za Dengel and Susneyos.

Susneyos, who became a determined Catholic, attempted to convert the country by force, and, guided in part by Paes's bigoted successor Alfonso Mendes, tried to eradicate many of the country's most venerated customs overnight. This led to strong popular opposition, and before long to extensive rebellion and civil war. After much bloodshed the Emperor realised the impossibility of achieving his aims. He reinstated the Orthodox faith, and then abdicated, in 1632, in favour of his son Fasiladas, who thereupon banished the Jesuits from his capital, and later from the entire country.

The advent of the Jesuits, brief and unsuccessful as it was, had a major impact on Europe's perception of Ethiopia. Benefiting from Susneyos's patronage they travelled widely throughout the empire, and saw more of it than any previous foreign visitor. They produced a series of informative annual reports on their activities. Three missionaries, Manoel de Almeida, Jeronimo Lobo, and Manoel Barradas, also wrote detailed accounts of their travels, in which they had much to say about the country, its people and their economic, social, and cultural life. Pero Paes, making use of Ethiopian royal chronicles in Ge'ez, which were until that time unknown in Europe, also published a very creditable history of the country he and his colleagues had failed to convert.

Another major development which occurred in 1636, only a few years after the expulsion of the Jesuits, was the founding by Emperor Fasiladas of an entirely new Ethiopian capital. Located at Gondar, north of Lake Tana, it was situated in a region where his father Susneyos had earlier constructed a series of short-lived capitals. Three centuries of moving capitals thus came to an end. Gondar differed from previous such settlements in that it remained at least the nominal centre of Ethiopian government for over two centuries, until the second half of the nineteenth century, and was thus characterised by a period of relative continuity which led to steady urban growth.

One of the most important events in the city's history was the construction on the orders of Fasiladas of a great castle-like palace, which was erected with the help of an Indian architect. Later emperors built a succession of grand structures, and thereby created an extensive imperial compound, by far the most impressive in this part of Africa, and one which has remained a place of fascination - and photographic interest - to this day.

Gondar, a great centre of imperial government, was the recipient of a constant flow of chiefs, officials and messengers from all over the country, and was the place where imperial armies often assembled prior to their departure on campaign. The settlement soon emerged as the largest in the country, with perhaps a hundred thousand inhabitants.

The city was also a major commercial centre. It dealt in gold, ivory, civet musk and slaves from the rich countries of the south-west, and did a thriving import-export business through both the Sudan and Massawa. Much of this trade was carried out by Muslims, who represented a minority of the population, but included some of the wealthiest persons in the land. Many Muslims in and around the city also served as weavers, tent-makers and porters.

Gondar was at the same time a great Christian religious centre. The site of numerous churches it was the abode of both the Abun, or Patriarch, who as in the past came from Egypt, and the Echagé, a native Ethiopian, who served as head of the monks. Many of the city's clergy were extremely learned. Some were able not only to recite the entire Bible by heart, but also to offer alternative interpretations of many of its passages. The church, being the principal patron of the arts, the settlement was likewise a major artistic and cultural centre, and a place where a new, and more representational, style of Ethiopian painting developed.

The city was equally famous for its handicrafts, which were ranked among the finest in the land. These included beautiful articles produced for the emperors and their families, for the nobility, and for the Church. A significant proportion of the city's craftsmen were Falashas, or Judaic Ethiopians, many of whom were blacksmiths, weavers and potters, as well as masons, who build both palaces and churches. Persons bringing wood to the city belonged to a related minority group, the Qemants, who have been described as a Judaic-pagan peasantry.

Gondar all in all was thus characterised by a much more complex division of labour than the rest of the country. The city's population was very varied, and in addition to courtiers, traders, and priests, included lawyers, parchment-makers, book-binders, church-painters, makers of spears and swords, shoe-makers, jewellers, courtesanes and prostitutes. The period after the departure of the Jesuits, and the restoration of the Orthodox faith, marked a new era in Ethiopia's foreign relations. Fasiladas severed all relations with the West, and went so far as to order that any Catholic priest entering the country should be executed. This policy of isolation from Europe continued for many decades, and was only gradually, and hesitatingly, relaxed.

Diplomatic contacts with the East were, however, expanded. Missions, several led by a notable Armenian named Murad, who served three successive emperors, were despatched to Persia, the Mogul empire, and the Dutch Indies. An envoy from Muslim Yaman was likewise courteously received. To consolidate such relations several Ethiopian rulers despatched remarkable gifts, including elephants and zebras, and at least one rhinoceros. Such animals were regarded as objects of wonder in their countries of destination, where some were sketched by local or foreign artists, who thus preserved their likenesses for posterity.

Notwithstanding Ethiopia's isolation from the West, many Ethiopian Christians continued to travel on pilgrimage to the Holy Land, whence a few made their way, as in the past, to Rome.

The Ethiopian colony in Rome contributed significantly to European scholarship on their country. In 1649, a learned young German, Hiob Ludolf, who had already taught himself a little Ge'ez, travelled to the Eternal City in the hope of advancing his knowledge of the language. Hearing that there were a number of Ethiopians there he hastened to see them. He described his ensuing meeting with them in a Latin account translated into seventeenth century English: "I addressed them", he recalls, "and acquainted them how desirous I was to learn the Ethiopick language. They surround me, and wonder, and at length demand the Reasons; to which, being heard, they return the Answer: that that could not be done out of Ethiopia, for it was a thing of great Labour, and much Time; that there was indeed one Gregorius [Gorgoreyos] there, a very Learned Man (whom they showed me), but that he neither understood Latin nor Italian".

The German replied that he "desired that they would only resolve some Doubts", and "satisfy" his "Difficulties", for he had "already acquired the Rudiments of that Language". At this point Gorgoreyos, "understanding from his Companions what I desired, immediately runs in, and fetches a great Parchment Book, curiously writ, and bids me read...[when he did so] they could not abstain from Laughter, especially Gregorius... But when I went about to interpret, they turned their Laughter into Admiration, scarce believing that that Language, which seemed so difficult (as they said) to the Fathers of the Society [i.e. the Jesuits], who abode so long in Ethiopia could be learned without a Master".

This was the beginning of one of the most remarkable relationships in the history of scholarship. Ludolf, who was young enough to have been the Ethiopian's son, recalls: "I daily visited Gregorius: But at the beginning we did not converse; for he understood no European language besides the Portuguese, which then I had not learned. He was then beginning to learn the Italian: So we did a long time confer by an Interpreter and at last began to discourse imperfectly ourselves. Afterwards we spoke in the Ethiopick, which neither of us had ever spoke; for among the Habessins the Amharic Dialect is used in speaking, the Ethiopick only in writing. Concerning speaking the Ethiopick, I had not so much as dreamed. So that we were forced, that we might understand each other, to use a Tongue to which neither of us had been accustomed".

Gorgoreyos proved a valuable informant, for he was a man of scholarly integrity and one who knew his country and its customs well. He was thus a valuable source of linguistic, historical and cultural information, and added greatly to what Ludolf was able to glean from the writings of the Jesuits and other travellers.

With such help Ludolf published the first Ge'ez dictionary and grammar in 1661, a notable history, the *Historia Aethiopica*, in 1681, an extensive commentary thereon in 1694, and the first Amharic dictionary and grammar in 1698. These six works were so important, and made such an enduring influence on European scholarship, that their author has been called the Founder of Ethiopian Studies in Europe.

The writings of the Jesuits, despite the latter's expulsion, continued to have an impact in Europe for many decades. One of those so influenced was Samuel Johnson, the future English lexicographer, whose first commission, of five guineas, was the translation of a French version of Lobo's memoirs into English. He later made Ethiopia the setting, in 1759, of his famous novel *The Prince of Abyssinia. A Tale.* Its first page bore the title "The History of Rasselas, Prince of Abyssinia", for which reason it became immediately better known as *Rasselas.* The name was derived from that of Emperor Susneyos's favourite brother, Ras Se'ela Krestos, who had befriended the Jesuits.

Ethiopia soon afterwards attracted the attention of a Scottish laird, James Bruce, who conceived the ambition of "discovering" the source of the Nile. This had in fact been done by the Jesuits a century and a half earlier, but he obstinately refused to accept their testimony. He had moreover no idea that the concept of "discovery" was, as we would now say, entirely Eurocentric, for the area had of course been known by its local inhabitants since time immemorial.

Bruce landed at Massawa in 1769, and spent the next few years in the Ethiopian interior, mostly at Gondar. He delved deeply into the country's history and tradition, arranged for the copying of many Ethiopian manuscripts, and took copious notes of what he saw and heard. On eventually returning to Scotland, he found that many of the things he said about Ethiopia were widely disbelieved. Asked at a dinner party whether the country had any musical instruments, he replied that he remembered seeing a lyre, whereupon the wit George Selwyn M.P. whispered to a neighbour, "I am sure there is one less since he came out of the country".

Undeterred by such scepticism Bruce, after much delay, decided to publish his memoirs. Unwilling to put pen to paper himself he dictated them to a secretary, and thus produced an at times pompous, opinionated, and gossipy five-volume, *Travels to Discover the Source of the Nile*, which appeared in 1790. Despite its failings it remains one of the most notable travel books ever written, and was immediately translated into French, and later often reprinted, in full or abridged form.

Bruce's *Travels*, were important in giving the European reading public glimpses of the Gondarine empire, which was then beginning to decline. He also introduced his readers to the Ethiopian royal chronicles, which, up to that point, had been available to the outside world only through Jesuit writings in Portuguese. His visit was also remarkable in that he brought back to Europe a copy of the Old Testament Book of Enoch, then unknown outside Ethiopia, and presented it to the library of King Louis XV of France. This opened up a new era in Scriptural studies, and revealed once again Ethiopia's abidingly important position in the Judaeo-Christian world.

Though Bruce was by far the most significant traveller of his time he was preceded shortly earlier by two other visitors: a Czech Franciscan missionary, Remedius Prutky, and an Armenian jeweller, Yohannes Tovmachean. Both visited Gondar, and wrote valuable accounts of the country. A French physician, Charles Poncet, who travelled there at the close of the seventeenth century, and treated Emperor Fasiladas's grandson, Iyasu I, the greatest of the Gondarine rulers, for a troublesome skin complaint, had earlier also written a useful memoir of his visit.

The decay of the Gondar-based empire which Bruce had witnessed, continued for over a century. Power, hitherto exercised by the emperors, was usurped by several provincial rulers, notably those of Tegray, who had easier access to firearms than their rivals in other parts of the country. None of the provincial nobles, however, were able to make themselves masters of the entire empire. They therefore spent much of their time fighting unsuccessfully among themselves. This period was described by Ethiopian authors as the Era of the Masafent, or Judges, for it resembled the time, described in the Biblical Book of Judges, when "there was no king in Israel: every man did that which was right in his own eyes". Increasingly important, but to the outside world then virtually unknown, kingdoms were at the same time emerging in the Muslim and animist lands to the south, east and west of the empire.

Disunity and civil war in northern Ethiopia continued into the second half of the nineteenth century. This state of affairs placed the country more and more at a disadvantage in relation to both Europe, which was by then rapidly becoming industrialised, and Egypt, which, due to its adoption of European technology and armament, was expanding southwards into the Sudan, and thence into Ethiopia's borderlands. Disunity and lack of strong government delayed Ethiopia's modernisation, which otherwise would probably have taken place at this time when the Industrial Revolution was beginning to spread from Europe to parts of Africa and Asia.

This period of disunity and civil war coincided with the arrival in the early nineteenth century of an increasing number of European travellers. Many of them wrote extensive, and in sometimes remarkably detailed, accounts, in not a few cases enlivened with revealing and remarkably fine engravings.[2] By such writings - and engravings - these travellers vastly expanded external perceptions of the country. The first of these writers came from Britain, but others soon arrived from France, Germany, Switzerland, Italy, and several other European countries.

Most of these travellers spent their time mainly in what is now northern and/or central Ethiopia. Visitors to Tegray included the Englishman Henry Salt, famous for his engravings, his loyal dependant Nathaniel Pearce, and a skilled amateur artist, Mansfield Parkyns, who married an Ethiopian wife. Then there was the meticulous German scientist Edouard Rüppell, an ambitious Belgian consul, Edouard Blondeel, the Swiss Protestant missionary Samuel Gobat, the French Saint Simonian missionaries Edmonde Combes and Maurice Tamisier, and their compatriots in a French scientific mission headed by Théophile Lefevbre. Among travellers to Shawa mention may be made of the British envoy William Cornwallis Harris, his compatriot Charles Johnston, who was a ship's surgeon, the French adventurer Rochet d' Héricourt, who paid three visits to the area, two German missionaries Karl Wilhelm Isenberg and Johann Ludwig Krapf, and their artist compatriot Johann Martin Bernatz. Gojjam was visited by the redoubtable British geographer Charles Tilstone Beke; Gondar by the missionary Gobat, and by the above-said French travellers to Tegray; Wallo by Isenberg and Krapf; and Harar by the indomitable British traveller Richard Burton, who spent ten brief but memorable days in the city.

The period of disunity, which continued into the second half of the nineteenth century, concluded with the rise of three important provincial rulers, Téwodros (reigned 1855-1868), Yohannes (reigned 1872-1889), and Menilek (reigned as Emperor 1889-1913), who successively achieved imperial power. It was their reigns, as we shall see, which witnessed the coming, and first flowering, of photography in the country.

Each of these monarchs attempted in their different way to unify and modernise their age-old country, and thus to create a new and more vibrant image of Ethiopia. Each ruler succeeded in some measure, but was to a greater or lesser extent frustrated, either by almost insuperable local difficulties or by the untoward interference of foreign powers.

Téwodros, whose power base was in the north-west of the country, where firearms were hard to come by, and faced strong opposition from both the Church and the provincial nobility, throughout his reign faced almost insuperable difficulties. His desire for unification and modernisation, which found expression in his yearning to import foreign craftsmen, led directly to his conflict with the British Government, and to his detention of a British consul, and later of a special envoy from Queen Victoria. This action led in turn to the despatch of a British expedition, which, after crossing most of northern Ethiopia without encountering any resistance, stormed his mountain citadel at Maqdala, in April 1868. Téwodros thereupon committed suicide rather than fall into the hands of his enemies. His efforts at unification and modernisation thus came to nought. The British expedition, which looted his capital, and took many treasures, including illustrated Ge'ez manuscripts, to London, was of no small photographic interest, for it took the first still extant photographs of the country. Téwodros's orphaned son Alamayahu, who was brought back to Britain by the victorious army, was, interestingly enough, the first Ethiopian of royal descent to be photographed.

Yohannes, a loyal son of the Church, who, unlike Téwodros, was prepared to temporise with the provincial nobility, was a more successful unifier than his predecessor, but also encountered immense difficulties with foreign powers. The first Ethiopian ruler to be photographed, albeit rarely, he resisted the Egyptians, who were then attempting to create an East African empire, in part at Ethiopia's expense. Despite their immense superiority in weapons, he defeated them twice, at the battles of Gundat and Gura, in 1875 and 1876 respectively. He had later to contend with armed penetration by the Italians at the coast, whom his commander, Ras Alula, thwarted at an encounter at Dogali early in 1887. After this an Italian author, Achille Bizzoni, rhetorically asked, "Who has ever heard of victories over the Abyssinians?" Yohannes was almost immediately afterwards confronted by an invasion by the Sudanese Dervishes to the west. He defeated these too, at the battle of Matamma in 1889, but at the close of the victorious engagement fell victim to a sniper's bullet, the last crowned head in the world to die in battle.

Menilek, a particularly photogenic figure, succeeded in obtaining large quantities of weapons which assisted him in unifying the greater part of the country, and eventually in establishing most of Ethiopia's pre-1935 frontiers. He nevertheless faced major foreign difficulties, notably with the Italians. In 1889, shortly after founding Ethiopia's modern capital, Addis Ababa, he signed a friendship treaty, the Treaty of Wuchalé, with them, only to find them using it, quite illigitimately, to claim a Protectorate over his country. He strenuously rejected this pretension, and declared that his country had no need of foreign protection, for Ethiopia, as prophesied in the Bible, "stretched forth her hands to God". The Italians advanced southwards to occupy most of Tegray, but Menilek mobilised, and defeated them in 1896 at the battle of Adwa, perhaps the greatest victory of an African over a European state since the time of Hannibal. This triumph shocked the European colonial powers then involved in the Scramble for Africa, who nevertheless flocked to open legations in Addis Ababa. The battle of Adwa was in fact important, for the country as a result of it survived the European Scramble. Ethiopia was the only state on the so-called Dark Continent to do so. This created for the country a new international image, particularly in the eyes of peoples of African descent: the image of an age-old African empire which had withstood the onslaught of European imperialism, and was thus a unique symbol of African dignity and independence.

Menilek then, at the very end of the nineteenth century and the beginning of the twentieth, began in earnest the difficult task of modernisation. To him goes the credit, as some of the photographs in this volume show, of introducing many innovations. These included the first coined money since ancient times, the first hospital and modern schools, the first postage stamps, the first bank, paved roads, and the railway between Addis Ababa and Jibuti. These innovations were largely carried out with the help of the three neighbouring colonial powers, Britain, France and Italy, which Menilek for a time played off, one against the other. Later, however, they joined together without his knowledge, and by the Tripartite Treaty of 1906 partitioned the country into three spheres of interest. On being subsequently told of this, he ironically thanked the representatives of the three colonial powers, but declared the agreement could not in any way bind his action.

The old monarch died in 1913. He was succeeded by his grandson Lej Iyasu, another highly photogenic figure, who, as we shall see, incurred the displeasure of both the Shawan nobility and the Ethiopian Orthodox Church - and was deposed three years later, partly, some claim, with the help of photography. He was succeeded by Menilek's daughter Zawditu, who was proclaimed Empress, while Tafari Makonnen (the future Emperor Haile Sellassie), who was the son of Menilek's cousin Ras Makonnen, was designated as Heir to the Throne. A difficult period of dual government followed, in which Menilek's policy of modernisation was nevertheless continued. Tafari, who was particularly interested in his country's international image, obtained Ethiopia's entry into the League of Nations in 1923, and in the following year embarked on a much publicised European tour.

Haile Sellassie, one of the most photogenic international rulers of his time, assumed the imperial throne in 1930, after which the pace of modernisation, as illustrated in this volume, was for a short time accelerated. The country's independence was, however, soon once again challenged. In 1935 the Italian fascist dictator Mussolini, seeking revenge for his country's defeat at Adwa, and wanting to seize for Italy a "place in the sun", launched an unprovoked invasion. The League of Nations branded fascist Italy as the aggressor, and imposed mild economic sanctions, but failed to halt the invasion - which lies outside the scope of the present photographic work. Fascist Italy, taking advantage of its overwhelming superiority in weapons, and aircraft, and making extensive use of mustard gas, succeeded in capturing Addis Ababa in May 1936, after which Mussolini proclaimed the creation of a fascist empire.

The Emperor later travelled to Geneva to appeal to the League of Nations, but did so in vain. The haunting figure of the frail monarch standing erect at Geneva, condemning the fascist use of poison gas against his people, nevertheless created yet a new image of Ethiopia. Many people throughout the world felt that failure to save Ethiopia had resulted in no less than the demise of the League. Meanwhile in Ethiopia itself many Patriots continued to resist throughout the ensuing occupation.

Mussolini's entry into the European war in 1940, we may conclude, transformed the situation, for it almost automatically brought Britain to the aid of the Ethiopian Patriots. This led in turn to the country's rapid liberation from fascist rule. Ethiopia, the first victim of fascist aggression, by a strange coincidence, thus became the first to be freed from Axis domination.

1 The Ge'ez alphabet is derived from the Sabaean, which it closely resembles. Ge'ez differs from the latter, however, in two important respects. 1) Sabaean had been written, like most Semitic languages, from right to left, or more often, in the "ploughshare" manner, in which the first line ran from right to left, the second from left to right, the third from right to left, and so on. Ge'ez, perhaps influenced by Greek from which Biblical texts were then being translated, was by contrast written exclusively from left to right. 2) Sabaean, like other early Semitic languages, made no use of vowels. Ge'ez by contrast employed small symbols attached to the main letter to signify them. The earlier alphabet was thereby transformed into a syllabary. Several Sabaean characters were at the same time modified in Ge'ez, some for example were turned round, so that a symbol which had been vertical became horizontal.

2 R. Pankhurst and L. Ingrams, *Ethiopia Engraved. An Illustrated Catalogue of Engravings by Foreign Travellers from 1681 to 1900*, London, 1988.

THE COMING OF PHOTOGRAPHY

OPTICAL SCIENCE made its first impact on the recording of Ethiopian images in the late eighteenth century, when the Scottish traveller James Bruce obtained a camera obscura. A cumbersome affair, it was constructed according to his specifications by a London firm, and was both "large and expensive". Hexagonal in shape, and six feet in diameter, it comprised two separate units, which folded together. The draftsman sat within, and drew without being seen. The instrument was valuable, he claims, for it enabled a person with limited drawing skill to draw buildings with rapidity, and achieve better work in an hour than the finest draftsman could, without it, do in seven. The resultant picture had the "inestimable advantage" of being "real", rather than imaginary, and was so detailed that passing clouds, and even the folds of peoples' dresses, could be indicated.

Despite Bruce's enthusiasm for the camera obscura in drawing buildings only one such Ethiopian picture is extant: a sketch of the largest standing Aksum obelisk. His other Ethiopian pictures consist principally of drawings of flora and fauna, and a few portraits, including those of two Ethiopian ladies of rank, Astér and Takla Maryam, subsequently published in J. M. Reid's *Traveller Extraordinary*. (London, 1968). It is sad that Bruce, one of the first foreigners to reside in the then capital, Gondar, left no pictorial image of it, or of its political, social and religious life. It should also be mentioned that virtually all the Ethiopian drawings the Scotsman claims to have made were in fact the work of his Italian assistant, Luigi Balugani, whose artistic achievements were recognised two centuries later, in Paul Hulton, F. Nigel Hepper and Ib Friis, *Luigi Balugani's Drawings of African Plants* (Rotterdam, 1991).

1. "Woodage Asahel", an 18th century Oromo chief: a drawing apparently based on James Bruce's camera obscura.

The earliest use of the photographic camera in Ethiopia dates back to the second half of the nineteenth century, which witnessed increasing contacts with Europe, as well as major advances in photography.

The first photographer to visit the country was Henry Stern, a British Protestant missionary of German Jewish descent, who arrived in 1859, to convert the Falashas, or Beta Esra'él people, to Christianity. Keenly interested in photography, he took many shots of people and places. These pictures have disappeared, but twenty were published as engravings in his book *Wanderings among the Falashas in Abyssinia* (London, 1852). They included a portrait of Abuna Salama, the Egyptian patriarch of the Ethiopian church; scenes of Emperor Téwodros's capital Dabra Tabor; a Falasha village; a Gondar palace; and a group of noblewomen with their slaves. Several other engravings based on the missionary's photographs, later appeared in A. A. Isaacs' *Biography of Stern* (London, 1886). Stern's photographs, though presented only through engravings, constituted the first camera-based documentation on the country, and thus inspire a sense of veracity lacking in drawings.

Stern, an opinionated bigot, incurred the wrath of Emperor Téwodros, who had him arrested, and his house searched. The officials charged with this task displayed much curiosity about the photographs, and, Stern claims, "entirely forgot their commission". They nevertheless carried off his photographic equipment, after which the Emperor enquired closely about the missionary, the illustrations in his book, and "the mode and method of taking photographs". Stern's Ethiopian servant Joséf, who was "supposed to be initiated in all the mysteries" of

2. Abba Salama, the Coptic Patriarch of the Ethiopian Church: an engraving based on one of the earliest photographs ever taken in Ethiopia, by the Protestant missionary Henry Stern, in the early 1860s.

the camera, if we can believe the German's ironic account, then gave the monarch a "most elaborate, and no doubt, most lucid explanation".

Stern's imprisonment, and that of several other Europeans, among them a British consul, Duncan Cameron, and Queen Victoria's envoy, Hormuzd Rassam, provoked the British Government, as we have seen, to despatch an armed expedition against Téwodros and his fortress at Maqdala, where the captives were detained.

The British expedition of 1867-8 made extensive use of the camera. The official history, Trevenen Holland and Henry Hozier's *Record of the Expedition to Abyssinia* (London, 1870), relates that shortly before the beginning of operations, the British Commander-in-Chief, the Duke of Cambridge, proposed that the Royal Engineers should provide a team to photograph sketches and plans made by staff officers. Seven photographers from the 10th Company of the Royal Engineers were accordingly attached to the expedition, and the then considerable sum of £477.6s.9d. sanctioned to purchase photographic equipment and materials. They were required primarily for photographing officers' sketches, and reproducing rapidly. Because of the mountainous terrain, and the absence of roads, it was decided that everything be packed in containers weighing not more than 40 kilos, and therefore suitable for transportation on mule-back. The camera selected was an 11 X 8in. sliding and folding instrument, but, being unsuitable for portraits, a Dallmeyer triplet lens for a 12 X 10in. plate was also purchased. It could take portraits a few feet from the sitter, or groups 20 to 30 feet away. The photographers' supplies, packed in 18 cases, included a photographic tent, or dark-room, a copying table, a printing frame, a portable still, a mounting frame, and a stock of chemicals and sensitised paper. During the advance inland one of the photographers fell ill at Sanafé, and on reaching Adegrat some chemicals ran short, and a telegraphic request had to be despatched for further supplies. Despite such difficulties the team made no fewer than 15,000 prints.

The expedition's photographs were important in that they were the first taken in Ethiopia by professional cameramen. Unfortunately for the country's cultural history their labours were, however, devoted mainly to matters of only ephemeral military interest. Photographs of historical or cultural significance received low priority. The expedition moreover by-passed Massawa, the principal port, Adwa, the main emporium of the north, and the ancient city of

3. Some of the troops of the 1867-8 British expedition, whose photographic unit took the first extant photographs of Ethiopia.

Aksum, and failed to reach such historical centres as Lalibala and Gondar. Furthermore no attempt was made to utilise the photographic team in making any scientific record of the areas visited. The Engineers did, however, take some photographs of Ethiopian personalities, settlements, and scenery, which have been preserved in the British Army Museum's Ogilby Trust and in the National Army Museum, both in London, and in the Royal Engineers' Institution, in Chatham, Kent. Several photographic albums on the expedition are also extant: one of the most complete, containing 65 pictures, in the Institute of Ethiopian Studies, in Addis Ababa.

Some of the expedition's most interesting pictures are a portrait of Emperor Téwodros's son, Prince Alamayahu, who was later taken to England, and of Mastewat, the Oromo queen of Wallo, with her eldest son Imam Ahmad, and a leading courtier. There were also group photographs of some Tegray chieftains, with Marcha Warqé, a young Ethiopian who had studied in India; the British commander, Sir Robert Napier and his staff; and Téwodros's European captives. Other pictures included views of Annesley Bay, or Zulla, where the expedition landed, scenes of the inland settlements of Sanafé, Antalo and Adegrat, and interior and exterior views of Maqdala, among them of its treasury and of its church, with the Emperor's dug grave nearby. There were also photographs of an Ethiopian minstrel with a one-string *masenqo*, or fiddle; a British army camp, and views of the rugged country the soldiers had to cross. Also of historical interest are photographs of two officer's sketches: of Dajazmach Kasa, the future Emperor Yohannes IV, and of Emperor Téwodros, immediately after his suicide.

Many of these pictures were later used to enrich accounts of the missionaries' detention, and of the expedition which secured their release. Several photographs appeared as engravings, in Stern's autobiographical memoir, *The Captive Missionary* (London, 1869), Roger Acton's *The Abyssinian Expedition* (London, 1872), and, much later, Percy Arnold's *Prelude to Magdala* (London, 1992). Actual photographs were also reproduced in F. Myatt's *March to Magdala* (London, 1970) and Darrell Bates's *The Abyssinian Difficulty* (London, 1979).

Prince Alamayahu's arrival in England also had photographic consequences. The orphan's portrait was taken by Julia Cameron, Britain's first professional woman photographer, who also photographed him with his British guardian, Captain Speedy.

4. Dajazmach Kasa, the future Emperor Yohannes IV: photograph taken by the British photographic unit of a sketch by William Simpson, an artist attached to the Napier expedition.

Alamayahu, on going to school in England, was also photographed by studio artists in Eton. He was the first Ethiopian royal to be immortalised by the camera, and photographs of him are preserved in the British Public Record Office.

In the decade after Téwodros's death another Ethiopian ruler, King Menilek of Shawa, arranged to employ three Swiss craftsmen. The most notable Alfred Ilg, a graduate of the Zurich polytechnic, arrived in 1879, and served as both technician and diplomatic adviser. An enthusiastic photographer he took many pictures with a large camera, but, as he recounts, was soon summoned to his master's presence. "I have heard something about you", Menilek declared, "which was very bad, and which I would not have believed you capable of. It is so ridiculous, so improbable, that I would not have believed it had I not heard it from trustworthy people". Ilg inquired what was wrong, whereupon Menilek replied, "I have been told that without my knowledge you made me very small, and stuffed me into a black box, with my whole town, houses, people and mules. And, what is even more unbelievable, I was standing on my head, with my legs in the air." Ilg was obliged, his biographer Conrad Keller

5. Portrait of Emperor Téwodros's son Alamayahu, taken in London around 1868 by Julia Cameron, one of the first important British women photographers.

relates, to explain the laws of optics, which "required some pedagogical skill", but passed off all right, so that Menilek soon "understood the workings of a photographic camera".[1]

Most of Ilg's photographs have apparently not been preserved, but some, dating from the late 1890s, are in the possession of his descendants in Zurich. Some have been reproduced by his second biographer, Willi Loepfe, in his *Alfred Ilg und die äthiopische Eisenbahn* (Frauenfeld and Zurich, 1978).

The last two decades of the nineteenth century witnessed the coming to Ethiopia and the Horn of several dozen European traveller-photographers, who, working first at the coast and later in the interior, amassed a wealth of photographic documentation. This development coincided with major advances in the printing art, which enabled photographs to be reproduced with increasing accuracy. The result was that engravings based on photographs were soon replaced by actual photographic reproductions. Travel literature thus underwent a great transformation, which provides the background to this volume. [2]

One of the first foreign travellers in the region to make extensive use of photography was the French geographer Georges Revoil, who visited northern Somaliland in 1880-1. He was followed by an Englishman, Frank James, who advanced from the northern Somali coast into the Ogaden in 1884-5. He recalls in his *Unknown Horn of Africa* (London, 1888) that the camera was then disliked by the "natives", for though some made no objection, "many were frightened and ran away". In one place he encountered "considerable abuse" from the inhabitants, who angrily "called all who sat for their portraits dogs and rats". Despite such opposition he published numerous photographs of the population.

The Somali region, and Ogaden, was also visited at this same time by another British traveller, Major Henry Swayne, who undertook the first of many journeys to the area in 1884. He observes in his *Seventeen Trips through Somaliland* (London, 1903) that he always kept a camera by his bed to capture any interesting situation that might arise. Urging the usefulness of small hand cameras, which had recently come on the market, he declared them "invaluable", and added: "I suggest that no large camera be used, nor chemicals, but that small photographs be taken with the hand camera and developed and enlarged on return to England". His book contains photographs of Somalis and their country taken by himself, and by a Polish traveller Prince Boris Czetwertynski.

Italy's seizure of Massawa in 1885 had led meanwhile to the arrival there of several Italian photographers, among them Luigi Fiorillo, Mauro Ledru, Francesco Nicotra, and Luigi Naretti, a cousin of Giacomo Naretti, a craftsman in Emperor Yohannes IV's service. They took pictures of the port and its inhabitants, and of the former Egyptian governor's palace. Subsequent Italian penetration, which was temporarily checked by Ras Alula at the battle of Dogali in 1887, was also accompanied by much photography.[3]

The advent of the camera in the Ethiopian interior was symbolised by the photographing of Emperor Yohannes, the first Ethiopian imperial ruler to be thus recorded for posterity. A picture of him with his son Ras Araya Sellasé, reproduced in this volume, is included in the French edition of Gabra Sellasé's chronicle of Menilek, *Chronique du règne de Ménélik II, roi des rois d'Ethiopie* (Paris, 1930-31). It also contains portraits, several by Naretti, of a number of nobles, among them the emperor's son, Ras Mangasha, and views of Adegrat, Maqalé, Adwa, and other northern cities.

Photography by this time was also taking place in Eastern Ethiopia. The Austrian ethnographer Dr Philipp Paulitschke, who travelled in the Somali and Oromo areas and Harar in 1885-7, published many photographs in *Harar* and *Beiträge zur Ethnographie und Anthropologie der Somal, Galla und Harari* (Leipzig, 1888). In the first he relates that the Amir of Harar's soldiers were happy to be immortalised in all their finery. Their master, Abdullahi, the city's last Amir, would not, however, agree to be taken until his *ullemas* were convinced that photography was not contrary to the Qoran. The Austrian explained that the Caliphs of Istanbul and Cairo and the Great Sheriff of Mecca had all been photographed, and that the camera did not produce any shadow like that made by graven images. The *ullemas* were at length persuaded, and the Amir's photograph was authorised. The chief nevertheless insisted that it should not be exposed in Harar, or indeed generally. Paulitschke also took a panoramic view of the city, and pictures of the houses of both nomads and settled coastal people. The Italian engineer Luigi Robecchi-Bricchetti, explored the Somali region and Harar, in 1888, and returned in 1890 and 1891. He took many photographs, including a portrait of Menilek's cousin, Ras Makonnen, the governor of the province, reproduced in his *Nell' Harrar* (Milan, 1896).

The first photographs of Ethiopia's western provinces were made at around the same time by a Frenchman, Jules Borelli, who visited Shawa and the Jimma and Kafa area in 1885-8. He took numerous photographs, including views of Menilek's palace at the then capital, Entotto, portraits, and scenery, reproduced in his *Ethiopie méridionale* (Paris, 1890).

The Italian Geographical Society had meanwhile established a station at Let Marafeya, near the old Shawan capital, Ankobar. The second director, Leopoldo Traversi, who arrived in 1888, was responsible for considerable photography. Pictures taken under his auspices included scenes of Ankobar and Let Marafeya, and several interesting portraits. Among them was one of Grazmach Joséf Negusé, who translated into Amharic the famous treaty of Wuchalé of 1889 on the basis of which the Italians soon claimed a Protectorate over Ethiopia; and another of Abba Jifar, ruler of Jimma. There was also a dramatic shot of Ras Mangasha bearing a stone in 1894 in submission to Menilek, who had by then assumed the imperial title. Several Traversi pictures were later reproduced as engravings in Guglielmo Massaia's *I miei trentacinque anni di missione nell' alta Etiopia* (Rome and Milan, 1885-1895); others later appeared as photographs in Traversi's Let-Marefià (Milan, 1931).

6. Amir Abdulahi of Harar, an engraving based on a photograph taken in the early 1880s by the Austrian ethnographer Philipp Paulitschke.

The southernmost stretch of Ethiopia had by then been investigated by a Hungarian, Count Teleki von Szek, who took photographs in 1888. Some were published by his companion Ludwig von Hohnel in *Ostäquatorial-Afrika* (Gotha, 1890) and *Zum Rudolf-See und Stephanie-See* (Wien, 1892), the latter translated as *Discovery of Lakes Rudolph and Stefanie* (London, 1894).

Travel and photography continued apace in the 1890s. An American traveller, Arthur Donaldson Smith, made his way across the Somali area to Lake Rudolf in 1894-5. He took photographs of the shrine of Shaikh Husein, but later discovered that his photographic plates had been soaked, and destroyed, in crossing the Ganana river. Luckily a young American companion, Fred Millett, had also taken a picture of the site, which is included in Smith's *Through Unknown African Countries* (London and New York, 1897).

In the northern Somali area meanwhile a British nobleman, Lord Wolverton, took photographs, in 1892-3, of the wild-life he had killed, which illustrates his *Five Months' Sport in Somaliland* (London, 1894). His compatriot Captain Sir Charles Melliss presented a not dissimilar collection in his *Lion-Hunting in Somali-Land* (London, 1895). So many such photographers had by then travelled in the Somali area that Evangelist de Larajasse's *Somali-English Dictionary* of 1897 lists a term for "photograph": the Arabic loan-word *sawir*, or picture.

The Italians, after establishing their colony of Eritrea in 1890, were meanwhile active in the north. Two Italian photographers, Roberto Gentile and Felice Scheibler, appeared in Asmara in the early 1890s. Eight photographic albums on the colony dating from this period are in the possession of the Biblioteca Reale in Turin. Though largely concerned with Asmara and Massawa they include scenes of the coastal settlements of Asab and Arkiko, the old monastery of Dabra Bizan, and the westerly towns of Keren and Agordat. There are also portraits of local chiefs, a variety of Eritrean "ethnic types", and some "native" dances. Several such photographs were subsequently reproduced in two commemorative volumes: Alessandro Triulzi's *L'Africa dall'immaginario alle immagini* (Turin, 1989) and Luigi Goglia's *Colonialismo e fotografia. Il caso italiano* (Messina, 1989).

At around this time an Englishman, James Theodore Bent, who had arrived in Eritrea in 1893, visited the antiquities in Tegray. He took many photographs of historical interest, including some of the first of the ancient obelisks of Aksum, a representation of a lioness cut on a rock, and murals from the church of Madhané Alam at Adwa, which are reproduced in his *Sacred City of the Ethiopians* (London, 1893).

In the following year a Frenchman, J. G. Vanderheym, landed at the newly established French Somaliland port of Jibuti, and proceeded to the still little known Ethiopian capital, Addis Ababa, founded less than a decade earlier. He recalls that Menilek, who had ascended the imperial throne only five years previously, "begged him to come to photograph him", and adds: "I spent a very interesting morning posing Empress Taytu, the princesses and the ladies of the court who had dressed in their most beautiful clothes. That day I also photographed Menilek's daughter", i.e. Princess Zawditu, who ascended the imperial throne a generation later. Vanderheym subsequently published many photographs in *Une expédition avec le négous Ménélik* (Paris, 1896).

Important Ethiopian personalities, as evident from the pictures in our volume, posed with the utmost care. They usually insisted on wearing official dress, and on adopting a dignified pose. Photographs, like the country's traditional paintings, tended to be neither informal nor relaxed. Group photographs, like the paintings, were also studiously composed. Persons were placed in hierarchical order, enabling the viewer to grasp everyone's relative importance, and relationship. Such photographs reflected Ethiopian norms, rather than the photographer's own values. Unrehearsed photographs, or "snaps", were still things of the future.

Italy's attempt to establish a Protectorate, and the subsequent fighting, which culminated in Menilek's victory at Adwa in 1896, led to intensified photographic activity by the invaders. Several Italian photographers took photographs of the Italian army, and of its "native" component. One such photographer was Michele Silvestri, who was attached to the Italian army, and later established an Asmara studio.

Italian photographers took pictures connected with the Adwa war: of the Italian soldiers who participated in it, the area in which it was fought, and contrived photographs of skeletons on the battlefield. A comprehensive collection of such shots was reproduced by an Italian journalist, Eduardo Ximenes, in his *Sul campo Adua* (Milan, 1897). Engravings based on photographs also found their way into Achille Bizzoni's *Eritrea nel passato e nel presente* (Milan, 1897), which was banned in the Italian colony. Rosalie Pianavia-Vivaldi's *Tre anni in Eritrea* likewise reproduces a photograph of Batha Hagos, who led a rebellion against Italian colonial rule in 1894. Several such pictures later appeared in Ridolfo Mazzucconi's *La giornato di Adua* (1896) (Milan, 1935), which included a portrait of Ras Alula. Another selection of such pictures was published in Roberto Battaglia's *La prima guerra d'Africa* (Roma, 1958).

7. After the battle of Adwa of 1 March 1896: an Italian photograph with doubtlessly rearranged skeletons.

An important consequence of Menilek's victory was that the number of foreign traveller-photographers arriving in Addis Ababa substantially increased, as the Great Powers despatched missions to the Emperor and established legations at his capital.

The year after the battle the British Government sent a mission led by Lord Rennell of Rodd, which included three amateur photographers, Lord Edward Cecil, H. H. Pinching Bey, and Major Swayne. Travelling by way of Harar they called upon Ras Makonnen, to whom they handed a photograph of the Coptic Patriarch of Alexandria, which he "received with much reverence". They also showed him photographs of Indian elephants and tigers. Swayne recalls that he and his colleagues "spent a long time" displaying such photographs, and that the

8. An early 19th century photograph of the Menilek palace compound, showing artillery captured at the battle of Adwa. Note recently planted, still slender, eucalyptus trees.

Ras was "much struck" by Indian elephants, and the manner in which they had been tamed in the sub-continent. Rodd and his party took numerous pictures, reproduced in Gleichen's *With the Mission to Menelik:* 1897 (London, 1898). Photography, and the gift of photographs, had become an adjunct to diplomacy, and when a subsequent British envoy, Captain John Harrington, presented Empress Taytu with Queen Victoria's portrait she rose from her seat, and, to his surprise, bowed deeply.

A French mission, led by Count Henri of Orleans, also arrived in Addis Ababa in 1897, and took many photographs, which appear in his *Une visite à l' empereur Ménélick* (Paris, 1898). It contains interesting pictures of the church and market at Harar, and court ceremonial, and religious and commercial activity at the capital.

A later mission from France headed by a French colonial administrator, the Marquis de Bonchamps, visited Ethiopia in 1897-8, when, journeying as far as the White Nile, it collected much photographic data. The second in command, Charles Michel, published numerous photographs, including a portrait of Menilek, ruins of the then abandoned capital, Entotto, and scenes of the far west, in *Vers Fachoda* (Paris, 1900). An album of almost 200 photographs was later deposited with the Société de Géographie in Paris. Other pictures by Michel were subsequently presented by his daughter to the Institute of Ethiopian Studies.

A third French mission, which arrived from West Africa after the diplomatic confrontation with the British at Fashoda in 1898, also took many photographs, notably of recently erected in Addis Ababa buildings, the Baro river, and the Ethiopian commander Ras Tassama and his warriors. These pictures were reproduced in Jules Emily's *Mission Marchand* (Paris, 1913).

The establishment of a French Roman Catholic mission at Harar had led meanwhile to the taking of photographs of the mission station, its leprosarium, dispensary and school, as well as notable missionaries, among them Monseigneurs André Jarousseau and Taurin Cahagne, and Ethiopian friends of the mission. These included the historian Alaqa Asmé, and Ras Makonnen's son Tafari, the future Emperor Haile Sellassie, as a child. These appeared in Martial de Salviac's *Les Galla* (Paris, 1900).

Several British travellers also arrived in the last years of the century. An intrepid military officer, Captain Montague Wellby, journeyed southwards to the borders of British East Africa in 1898-9, and took shots of Ethiopian military exercises and scenes in the south, and of the palaces at Addis Ababa and

9. The early twentieth century Ethiopian historian Alaqa Asme and his family.

Harar, which were published in *'Twixt Sirdar and Menelik* (London, 1901).

Herbert Vivian, who made his way from Zayla to Addis Ababa in 1899-1900, states in his memoirs, which bear the romantic title *Abyssinia. Through the Lion-Land to the Court of the Lion of Judah* (London, 1901), that strong popular opposition to photography still existed. At Aden he had tried to snap a "delightful" Somali chambermaid, but hardly had he brought out his camera, when this "usually silent creature emitted a shrill scream, and fled". On reaching Harar he learnt that "strict orders had been given that the cathedral should not be visited except during the hours of service, lest it be desecrated by attempts at photography". Notwithstanding such difficulties he photographed Ras Makonnen's palace and the Harar mosque, and, on arriving at the capital, the Emperor and Empress, their palace, and the city's market, attended by thousands of people. He also took shots of the British Agency (later Legation), situated in a traditional wattle and daub house, and of the recently constructed church of Ragu'él at Entotto, an octagonal building designed by an Indian, Haji Khawas. Other pictures include a sequence depicting the construction of a traditional hut, and the custom of chaining a plaintiff and defendant together.

Widespread reluctance to be photographed was also noticed by the big-game hunter, Percy Powell-Cotton. He recalls in his *Sporting Trip through Abyssinia* (London, 1902) that while photographing in the Addis Ababa market he aroused "a good deal of curiosity", and one girl, learning that he was taking her portrait, ducked, and exclaimed, "By the Holy Trinity, tell me, what is he doing?." He later proceeded to Gondar, where he photographed the town and its palaces. These and other pictures are preserved in England at the Powell-Cotton Museum at Quex Park, near Birchington, Kent.

For other pictures of the time we are indebted to another British big-game hunter, Sir Alfred Pease. A keen photographer he took many shots, which he refers to as "the amateur produce" of his Kodak camera. He published photogravure reproductions of them in his *Travel and Sport in Africa* (London, 1902), which contains interesting market scenes at Addis Ababa and Harar. Yet another photographer of the time was a Polish nobleman, Count Joseph Potocki, who reproduced pictures of Somali wild life in his *Sport in Somaliland* (London, 1900), also published in Polish.

The opening of the twentieth century witnessed the coming of increasing numbers of photographer-travellers. A Frenchman Hugues Le Roux, who went to the western province of Wallaga in 1900-1, took pictures of architectural developments in the principal towns, as well as pictures of everyday life, and published them in his *Ménélik et nous* (Paris, 1902). In his preface he draws attention, as a good Frenchman, to the fact that his camera was French, a vérascope Richard, which he describes as an instrument of "unique precision". It had enabled him to present the French Government, French geographical societies, French chambers of commerce, and the whole of France, with a hitherto unknown image of the "real" Abyssinia.

A British nobleman, Lord Hindlip, who explored the Rift Valley in 1902, reproduced photographs taken by himself and his wife in his *Sport and Travel* (London, 1906). His compatriot, Arthur Hayes, who travelled around Lake Tana in 1903, took a variety of unusual pictures, including a market scene on the Zegé peninsula, a child victim of leprosy, persons crossing the lake on a *tankwa*, or papyrus boat, and frescoes from the lake-side church at Qorata.

Emphasising the fascination evoked by the camera he reported, in *The Source of the Blue Nile* (London, 1905), that crowds gathered to look at his instrument which was still "an unknown marvel".

Two French scientific missions also arrived in the early years of the century. The first, led by Vicomte Robert du Bourg de Bozas, visited the southerly provinces of Balé, Arsi, Guragé and Sidama in 1901-2, and took photographs, of "ethnic types", which were published in its memoirs, *De la mer rouge à l'Atlantique* (Paris, 1906). The second mission, headed by Jean Duchesne-Fournet, journeyed westwards to Gojjam and Wallaga in 1901-3, and took numerous pictures, which are reproduced in its *Mission en Ethiopie* (Paris, 1909). It also contains photographs of costumes and noblemen's seals. Three albums containing almost a thousand of the mission's photographs are preserved by the Société de Géographie in Paris.

A German professor, Dr Emil Schoenfeld, who made a trip through Eritrea in 1903, likewise took a number of fine photographs, including pictures of the old palace at Massawa, the crowded "native market"

10. The arrival of the USA on the Ethiopian diplomatic scene: the American party photographed in front of the Menilek palace in Addis Ababa, 1903.

at Asmara, and new Italian buildings at Ghinda and Karan. They appeared in his *Erythräa und der ägyptische Sudan* (Berlin, 1904).

Several diplomatic and military missions also made their appearance in Addis Ababa at this time. The first embassy from the United States arrived in 1903-4. Its leader, Robert P. Skinner, reproduced many photographs in his *Abyssinia of To-day* (London, 1906). They included one of the venerable Emperor on his throne, the stout Empress Taytu, artillery captured from the Italians at Adwa, eucalyptus trees newly introduced from Australia, plantations of coffee, by then a major export, the ritual dance of the clergy, and the American mission.

A small British military team also came in 1903-4 to serve with Ethiopian forces operating against Muhammad ibn 'Abd Allah, the so-called Mad Mullah of Somaliland. It took photographs of Menilek's soldiers, some then being trained on modern lines, and views of Harar and Ogaden, which were published by Major James Jennings and Dr Christopher Addison in *With the Abyssinians in Somaliland* (London, 1905). Photographs of this and later campaigns against the Mullah, and of the Somali leader's encampment at Talé, appeared in Douglas Jardine's *Mad Mullah of Somaliland* (London, 1923), and, later, in Ray Beachey's *The Warrior Mullah. The Horn of Africa Aflame* 1892-1920 (London, 1990).

The first German mission to Ethiopia arrived in 1905, and returned by way of Gondar and Aksum. It took excellent photographs of these cities, church murals, market crowds, bands of warriors, contemporary paintings, musical instruments, and jewellery. Some of these appeared in Felix Rosen's *Eine deutsche Gesandtschaft in Abessinien* (Leipzig, 1907), and others in Hans Vollbrecht's *Im Reiches des Negus Negesti Menelik II* (Stuttgart, Berlin and Leipzig, 1906).

Even more important was the German archaeological mission led by Enno Littmann, which carried out research in the Aksum area in 1905-6, and published its findings in his *Deutsche Aksum-Expedition* (Berlin, 1913). The mission took photographs of the obelisks, inscriptions, ruins and coins of Aksum, as well as of the ancient temple at Yéha, the Aksumite dam at Kohaitu, the early church of Dabra Damo, and panoramic views of both Aksum and Adwa. Also of interest were portraits of the local governor, Dajazmach Gabra Sellasé, and various officials and priests.

Intensified interest in Ethiopia on the part of Italy in the years after the Adwa war led meanwhile to the appointment at the Italian Legation in Addis Ababa in 1901 of a physician, Dr Lincoln de Castro, who travelled extensively in the country for the next decade. He took numerous photographs, of historical sites, palaces, churches, houses, and markets; secular and religious ceremonies and festivals; and sports and games, including Ethiopian chess and the board-game *gabata*. He also photographed innovations introduced by Menilek, such as the telephone

11. Menilek's seven-year-old grandson Lej Iyasu, the future "uncrowned king", on his mule near Ankobar, photographed by a visiting Italian diplomat.

and the mint, and took portraits of prominent personalities, among them of Menilek's young grandson and future heir Lej Iyasu, and close-ups of ethnological objects, all of which he published in his *Nella terra dei Negus* (Milan, 1915).

Another photographer attached to the Italian Legation was a consular agent, Bertolani, who took over 300 photographs, now preserved in the Istituto Italo-Africano in Rome. They included portraits of Menilek; Abuna Matéwos, the Egyptian head of the Ethiopian church; and provincial governors, some of which are reproduced in Gabra Sellasé's chronicle. Extensive collections of photographs were also published by several other Italians, among them Ottorino Rosa, whose *L'impero del Leone di Giuda* (Brescia, 1913) contained over a hundred pictures, and Dr Carlo Annaratone whose *In Abissinia* (Rome, 1914) had almost two hundred. Also noteworthy was Arnoldo Cipola, a Corriere della Sera reporter who travelled from Eritrea to Addis Ababa in 1910, and published over 150 photographs in *Nell' impero di Menelik* (Milan, 1927) and more in later works. His pictures included many provincial scenes, among them Emperor Yohannes's palace at Maqalé and shots of Adwa and Harar. He also took portraits of Fitawrari Habta Giyorgis, Ras Tassama, and other members of Menilek's newly established Cabinet, and of the old Emperor's recently appointed heir Lej Iyasu. Other photographs showed patients at the Italian Legation hospital, and Addis Ababa's new institutions, among them the telegraph office, the Bank of Abyssinia, and the customs office.

Yet another photographer of this period was an Englishman, Arnold Savage Landor, who journeyed for the "sole object" of pleasure in 1906. On reaching Harar he took what he claims was the last photograph of Ras Makonnen before the latter's death a few weeks later, and one of the Ras's young son Tafari. Describing the difficulty of persuading them on to a balcony with sufficient light to photograph them, he recalls that he "could not help being amused" at their "great fear of the sun". In Addis Ababa the ailing Emperor granted him an audience, and agreed to let him take his portrait, but was willing to wear his heavy royal regalia for only one minute. The British envoy, Captain Harrington, who accompanied him into the royal presence, insisted on timing this watch in hand, with the result, Landor complains, that "when I came to take the photographs I took several on the same plate. Having discovered my mistake, I took others, but Menelik's head was shaking so violently with the effort of supporting the imperial emblem, that they, too, were not successful". Landor also took photographs of the Emperor without a crown, which, he claims, were "slightly better". Landor's pictures, reproduced in his *Across Widest Africa* (London, 1907), included one of Menilek watching by telescope the arrival of almost eight thousand guests for a state banquet. There was also a panoramic view of the hill upon which the palace was built, and scenes of the inland port of Gambela and of its tall dark-skinned inhabitants. Many photographs, including some of Addis Ababa, and its diplomatic round, were also taken by the British minister to Menilek's court, Wilfred Thesiger, who

12. Capturing an image: a French photographer at work around 1930. Note camera and photographer of the period.

arrived in 1909. They were subsequently deposited in the Pitt Rivers Museum in Oxford.

Bede Bentley, yet another British photographer, made his way from Jibuti to Addis Ababa in 1907-8 to bring Menilek a British car. Bentley took photographs of the vehicle's progress to the Ethiopian capital, where the Emperor was persuaded into the driver's seat, and was snapped laughing with pleasure. These photos appeared in Clifford Hallé's *To Menelik in a Motor-Car* (London, 1913), which also contains pictures of Harar's main gate, an early printing press, run by Lazarist missionaries, and one of the capital's first hotels, founded by Empress Taytu. Pictures of Bentley's trip were later also reproduced in T. R. Nicholson's *A Toy for the Lion* (London, 1965).

A car from Germany was shipped to Menilek almost immediately afterwards by a German traveller, Arnold Holtz, who published pictures of it, as well as portraits of the monarch, his spouse, and his palaces at Addis Ababa and nearby Holota or Gannat, in *Im Auto zu Kaiser Menelik* (Berlin, 1908). Holtz later took for study in Germany an Ethiopian student, Tassama Eshaté, who became his country's first amateur photographer: According to his grandson Tadele Yidnekachew, he possessed no fewer than fifteen cameras.

The coming of photography by this time had a culturally important impact on Ethiopian funeral procedure. Hitherto it had been customary to display an effigy of the deceased during the funeral procession, but in the case of persons for whom photographic portraits were available these came to be used instead. One of the first such occasions was the funeral of Ras Makonnen in 1906 when the priests of Addis Ababa's churches are reported to have remained all day beside his portrait and catafalque.

Despite the increasing importance of the camera no word for "photograph" was known to the European lexicographers of Amharic, the main language of central Ethiopia, at the turn of the century. In Eritrea use was, however, recorded of the term *sa'ali*, literally picture, in Francesco da Bassano's Tegrenya-Italian dictionary of 1918.

Traveller-photographers continued to arrive towards the close of Menilek's reign. Three scientists sent by the Italian Geographical Society, Giuseppe Ostini, Alfonso Tancredi and Maurizio Rava, journeyed to Eritrea and western Ethiopia in 1908, and took many pictures which can be seen in an album in the Istituto Italo-Africano in Rome. Several of these photographs, including views of Asmara, Aksum, Adwa, Gondar, the Samén mountains, and the Blue Nile falls, appeared in Rava's *Al lago Tana* (Rome, 1913). A German forester, George Escherich, meanwhile carried out two journeys, in 1907 and 1909, and, travelling south to Lake Rudolf, published photographs of the area in *Im Lande des Negus* (Berlin, 1921). Jacques Faitlovitch, who visited the Falashas in the north-west, published pictures of them, and of their handicrafts, in *Quer durch Abessinien* (Berlin, 1910), which also contains portraits of several educated in Europe. A British Indian army officer, Captain Arthur Mosse, who went game-hunting in British Somaliland and Ogaden, published photographs of wild animals, sadly for the most part slaughtered, in *My Somali Book* (London, 1913). Another British officer, Captain Chauncy Stigand, proceeding southwards to Lake Rudolf, reproduced pictures of the area, in *To Abyssinia: Through an Unknown Land* (London, 1910).

Photographic activity continued unabated throughout the rest of Menilek's reign. A Swiss geographer, Georges Montandon, travelled in the west of the country between 1909 and 1911, and took many pictures of anthropological and ethnographic subjects, which appear in his *Au pays Ghimirra* (Neuchâtel, 1913). Other photographs were taken at this time by an Italian, Carlo Citerni, who, travelling between Addis Ababa and Italian Somalia in 1910-11, published pictures of Amharas, Oromos, and Somalis in his *Ai confini meridionali dell' Etiopia* (Milan, 1913). It also contains a view of the Italian Legation in Addis Ababa, and of the capital's church of St George still surrounded by scaffolding. An album of his zincotypes is preserved in the Istituto Italo-Africano. Another Italian, Luigi Malvezzi, an engineer attached to the Agenzia Commerciale in Gondar in 1912-14, also took numerous photographs, likewise housed in that institute.

The quarter of a century between Revoil's travels of 1880-1 and Menilek's death in 1913 thus witnessed a steady expansion in photographic activity. Travellers, by then criss-crossing the country, were relying increasingly on the camera to embellish their writings, and in little more than a generation built up a comprehensive corpus of photographic documentation.

Menilek's death, and the succession of his young grandson, Lej Iyasu, was followed by the outbreak a year later of World War I. A power struggle, in which the legations of the warring European powers were closely interested, soon erupted, and the number of visiting photographers declined. Photography seems, however, to have risen to new heights of importance, for several politically significant photographs of Iyasu were made and published.[4] One of the first shows him as a child, but already wearing a lion's hair head-dress like that worn by warriors. Seated beside him is his tutor, the emperor's trusted courtier Ras Tassama, who held the honorific title of Bitwaddad, literally the Emperor's "Beloved". The chief places a fatherly hand over that of his ward, to show that Ethiopia's destiny was in tried, as well as youthful, hands. A similar message was projected in a painting in the Church of St Mary at Addis Alam, west of the capital. No less politically significant were photographs of Iyasu seated beside Menilek, emphasising that the prince was the old ruler's chosen heir.

Unsubstantiated report has it that an even more politically important photograph (or, some say, group of photographs) of Iyasu, was taken. To understand the background to the story it should be recalled that Iyasu spent much time in the Muslim periphery of the country, and that this alienated a number of Menilek's old supporters. Opposition to Iyasu came from many of the Shawan nobles, who feared that he would give undue influence to courtiers from his father Ras Mika'él's home province, Wallo. The legations of the Allied Powers, Britain, France and Italy, also suspected that the prince, who had pro-German affiliations, might side with the Central Powers. Opposition to the young ruler soon crystalized around the accusation that he had abandoned Christianity in favour of Islam, from which his father had earlier been forcibly weaned by Menilek.

A widely held Ethiopian tradition, which however is not corroborated by contemporary documentation, holds that the movement against Iyasu was fanned in Church circles by a photograph depicting him in Muslim dress. Sight of this picture, it is claimed, persuaded the Patriarch and the Echagé, or head of the monks, to free the nobles from their oath of allegiance. It is further asserted that this was not a genuine photograph, but a "doctored" picture, produced by a local Armenian photographer, Levon, or Léon, Yazegjian. Another version of the story holds that the photograph was the work of British Intelligence, while others go so far as to assert that pictures were distributed showing Iyasu "surrounded by Muslim priests at prayer", or "in a harem with dozens of white women", and bore an Amharic caption reading "The Anti-Christ". It is may be noted that fake photographs of Emperor Haile Sellassie in similarly compromising situations were distributed long afterwards, on the eve of his overthrow in 1974.

The mystery surrounding the alleged fake photograph, or photographs, of Lej Iyasu is compounded by the fact that at least two apparently genuine photographs of him in non-Christian garb were taken. One shows him in the Muslim city of Harar, wearing a turban. Standing beside him is a prominent city leader, Abdullahi Ali Sadeq, whose descendants testify to the veracity of the picture. The second shows him dressed as an Afar, or Danakil, with an Afar dagger at his waist. If these photographs are genuine, what purpose was there, it may be asked, of producing doctored ones?

Two other, authentic, politically significant photographs of Iyasu deserve mention. One, which depicts him wearing a cross, was published in 1915 in the Addis Ababa Amharic newspaper *Aimero*. This portrait, made, interestingly enough, by the said Yazegjian, was perhaps printed to reassure Christian fears as to Iyasu's alleged apostasy. The other shows the prince kneeling, in filial devotion, beside his father Ras Mika'él, who has his hand on his son's head, to demonstrate confidence in his offspring. The picture's message was further underlined by what would seem to be Menilek's crown, placed immediately behind the Ras.

After Lej Iyasu's overthrow in 1916, Menilek's daughter Zawditu was crowned Empress, and Ras Makonnen's son Tafari, Heir to the Throne. This inaugurated a difficult period of dual government, with power divided between two leaders, each with his or her own palace, court and government, and is symbolised in another politically interesting picture, in which the two rulers are seen seated apart.

The new political dispensation, which was supported by the three neighbouring colonial powers, opened the country once again to foreign travellers. Several gained access to Ras Tafari, who was in charge of foreign relations, and hence figures prominently in European writings of the period. A prominent foreign visitor of this time was a British businessman, Charles Rey, a keen photographer. His first book, *Unconquered Abyssinia* (London, 1923), contained portraits of Iyasu, Zawditu, and Tafari, as well as Addis Ababa market scenes, priests dancing, the cleaning of cotton, weaving, and basket-making. A later work, *In the Country of the Blue Nile* (London, 1927), was illustrated with a signed portrait of Tafari, and others of his spouse Wayzaro Manan, Ras Haylu, the ruler of Gojjam, and his daughter Sabla Wangel, wife of Lej Iyasu. Rey also photographed a local governor of Gudru, Fitawrari Ayalaw, who, like many chiefs of that time, was "very anxious to have a portrait of himself". He "despatched a horseman to his abode to bring out his full war regalia of lion's-mane collar and head-dress, embroidered cloak, embossed shield, and the war-trappings of his pony, to do justice to the occasion". His men then "formed a ring around him", while he "arrayed himself in his panoply, and thus prepared, invited us to immortalise him". Other shots depict the capital's horse races, a source of great interest among the foreign diplomatic community, Menilek's steam-roller, the newly opened Tafari Makonnen school, and a rural wedding.

Several British consular and other officials were meanwhile taking photographs in the provinces. Major Robert Cheesman, a consul in the north-west, wrote a scholarly work, *Lake Tana and the Blue Nile* (London, 1936), which was illustrated with pictures of its inhabitants and antiquities. His compatriot Major Arnold Hodson, consul in the south, produced two books. His first, *Seven Years in Southern Abyssinia* (London, 1927), contained a portrait of Abba Jifar, as well as shots of a Walamo grave, a traditional boat on Lake Abaya, and the British consulate at Mégga near the Kenya frontier. *His work Where Lion Reign* (London, n.d.) reproduced photographs of the Anuak, Maji, and Boma people. Major Hubert Maydon visited the north, and included scenes of its flora and fauna in his *Simen, its Heights and Abysses* (London, 1925), while Major Henry Darley, an officer at Mégga published pictures of slaves and slave-hunters in his *Slaves and Ivory* (London, 1926). Major Powell-Cotton returned to Ethiopia in 1924, and took further photographs, preserved in the museum bearing his name.

A British woman author, Rosita Forbes, meanwhile travelled via Harar to Addis Ababa with a ciné-cameraman, Harold Jones, and met both Zawditu and Tafari Makonnen, who appear, revealingly seated apart, as the frontispiece to her book *From Red Sea to Blue Nile* (London, 1925). It also includes photographs of the Addis Ababa market, and huge crowds waiting to accompany the Empress. Author and cameraman then proceeded northwards to photograph the rock-hewn churches of Lalibala, the Gondar palaces, and the obelisks of Aksum.

The most important French travellers of these years included Charles Michel-Côte, who visited Ethiopia in 1919-20, and deposited over 200 photographs in the Société de Géographie in Paris. Many photographs were also taken by the writer Jean d'Esme, whose book *A travers l'empire de Ménélik* (Paris, 1928) contains portraits of Tafari, described as "La lumière d'Ethiopie", and Ras Haylu of Gojjam, participating symbolically in agricultural work. Other photographs depict Addis Ababa scenes, including open-air hair-dressing and street sewing-machines, hippopotamus-hunting in the Awash river, and Dabra Berhan Sellasé church at Gondar, with wooden columns, which in the course of restoration were later removed. There is also a shot of an inquisitive countryman inspecting the traveller's camera and tripod. Photographs of Tafari Makonnen and his contemporaries likewise appeared in Pierre-Alype's *L'empire des négus* (Paris, 1928).

Also of interest were snaps taken by Henri Rebeaud, a French teacher at the Tafari Makonnen School, published in his *Chez le roi des rois d'Ethiopie* (Paris, 1934), which shows the school and its students. Other historically interesting pictures were taken by the French adventurer Henri de Monfreid whose photographs are housed in the Musée de l'Homme in Paris. Some are reproduced in his many publications, such as *Vers les terres hostiles de l'Ethiopie* (Paris, 1933), *Le lépreux* (Paris, 1935), *Le drame d'Ethiopie* (Paris, 1935), and *Le masque d'or ou le dernier négous* (Paris, 1936). They feature market and caravan scenes, chiefs, ritual dances, a lion-hunter, sorcerers, and such gruesome sights as amputations and public hangings.

Several Italian works also contained valuable photographic illustrations. Renzo Martinelli's *Sud* (Florence, 1930) featured Eritrean tribal figures, as well as views of Addis Ababa. Generoso Pucci's *Coi 'negadi' in Etiopia* (Florence, 1934) presented photographs of rivers and plants, among them the *dum* palm and the *baobab* tree, and views of the Gondar castles. The latter are also seen in Raffaele di Lauro's *Tre anni a Gondar* (Milan, 1936), which reproduced photographs of the bones of the nineteenth century British consul, Walter Plowden, the "queen of the zars", or spirit-possessed, and Dajazmach Wandwassan Kasa, a young nobleman who was soon to be executed by the Italian fascist invaders. Raimondo Franchetti's *Nella Dancàlia etiopica* (Verona, 1930) included interesting photographs of the Afar area, while Enrico Cerulli's *Etiopia occidentale* (Rome, 1930-3) contained a portrait of Abba Jifar and his wife Genné Mingitti, besides pictures of Oromo and Kafa dress, ritual dancing, panning for gold and platinum, and traditional bridges over the Gojjab river. Anglo-Italian scholarship led to the publication of Lewis Mancano Nesbitt's *La Dancalia esplorata* (Florence, 1930), which appeared in English as *Desert and Forest* (London, 1934) and *Hell-hole of Creation* (New York, 1935). They were illustrated with photographs of a salt caravan, the fording of the Awash river, and bare-breasted Afar women.

Prominent among German travellers was Hermann Norden, whose memoirs were published in *Durch Abessinien und Erythräa* (Berlin, 1930), translated as *Africa's Last Empire* (London, 1930). Based on extensive travels, they showed such varied items as the Menilek mausoleum in Addis Ababa, traditional Ethiopian painting, cotton and salt merchants at Gondar, and a Falasha priest reading his Ge'ez Bible. Another German, Max Grühl, visited Kafa, and published many photographs of it in his *Vom heiligen Nil ins Reich des Kaisergottes von Kaffa* (Berlin, 1929), translated as *The Citadel of Ethiopia. The Empire of the Divine Emperor* (London, 1932).

A no less important German contribution was made by Joseph Steinlehner, a professional photographer attached to the Afrika-Photo-Arkiv of Munich, who took many interesting shots. Mention must also be made of Kurt Lubinski, another professional, who took numerous photographs, reproduced in his *Hochzeitsreise nach Abessinien* (Leipzig, 1929), and its Dutch and Swedish editions, *Abessinië. Land en Volk* (Amsterdam, 1935) and *Abessinien* (Stockholm, 1936). They showed Tafari Makonnen School students, paintings by artist Belachaw Yimer, a German ostrich farm, a railway engine, and the camera Lubinski used.

Several American travel writers also published photographs. Edward A. Powell's *Beyond the Utmost Purple Rim* (New York, 1925) contained shots of Menilek's old palace at Holota, and Tafari's bodyguard. James E. Baum's *Savage Abyssinia* (London, 1928), reprinted as *Unknown Ethiopia* (New York, 1935) included a picture of a messenger carrying a letter in a traditional cleft stick. Gordon MacCreagh's *Last of Free Africa* (New York, 1928) showed a beggar travelling on horseback. The *National Geographic Magazine* for June 1930 published a well illustrated article, by Harry V. Harlan, featuring a mule journey via Ankobar and Dasé to the Lalibala churches.

Emperor Haile Sellassie's coronation in November 1930 was a notable media event, attended by numerous foreign journalists, and cameramen, besides important visiting dignitaries. This gave the country considerable coverage in pictorial journals, among them the *Illustrated London News*, in England, and *La France Illustrée* and *L'Illustration* in Paris. The June 1931 issue of the American *National Geographic Magazine* likewise published over a hundred photographs of Ethiopia, many in colour. This burst of photographic activity inspired the production half a century later of two commemorative photographic volumes Bertrand Hirsch and Michel Perret's *Ethiopie, année 30* (Paris, 1989) and Africa Archive's *Addis* 1930. *The Coronation of H.I.M. Emperor Haile Sellassie* (London, 1992).

One of the first books to appear after the coronation was the Comtesse de Jumilhac's *Ethiopie moderne* (Paris, 1933). It contained photographs of the new imperial family in their regal dress, the Duke of Gloucester and Maréchal Franchet d'Espéry, the British and French representatives to the coronation, a portrait of Shaikh Hojalé al-Hussein, the ruler of Bela Shangul on the Sudan frontier, and numerous scenes of Addis Ababa, Dire Dawa, and Harar. Photographs of these towns were also published in François Azaïs and Roger Chambard's *Cinq années de recherches archéologiques en Ethiopie* (Paris, 1931), which contains archaeological and ethnographic scenes of Soddo, Guragé, Sidamo, Konso, and Walamo, now housed in the Société de Géographie in Paris. Another French scholarly mission, which travelled from Dakar to Jibuti took no less than 2,500 Ethiopian photographs, among them shots of Gondar, spirit possession, and other lesser known aspects of cultural life, which were deposited in the Musée de l'Homme. A few were reproduced in Michel Leiris's *L'Afrique Fantome* (Paris, 1934), and in the June 1933 issue of *Minotaure*.

British photographic visitors of the time included Wilfred Thesiger, junior, who took pictures in the Afar region, now deposited in the Pitt Rivers Museum. Some also appeared in his memoirs *The Life of My Choice* (London, 1987). Two gold and platinum prospectors in the far west, Eustice Bartleet and Frank Hayter, published photographs of their work in their writings, respectively entitled In the *Land of Sheba* (Birmingham, 1934) and *In Quest of Sheba's Mines* (London, 1935). A Swiss, Werner Mittelholzer, reproduced aerial photographs, in his *Abessinienflug* (Zurich, 1934).

Topical photographs were likewise featured in the writings of two Americans. Catherine M. Jacoby's *On Special Mission to Abyssinia* (New York, 1933) published Addis Ababa street scenes, portraits of prominent officials, among them the foreign-educated Kantiba Gabru, and Haile Sellassie's coronation coach. John H. Shaw's *Ethiopia* (New York, 1936) contained many interesting shots of Addis Ababa buildings.

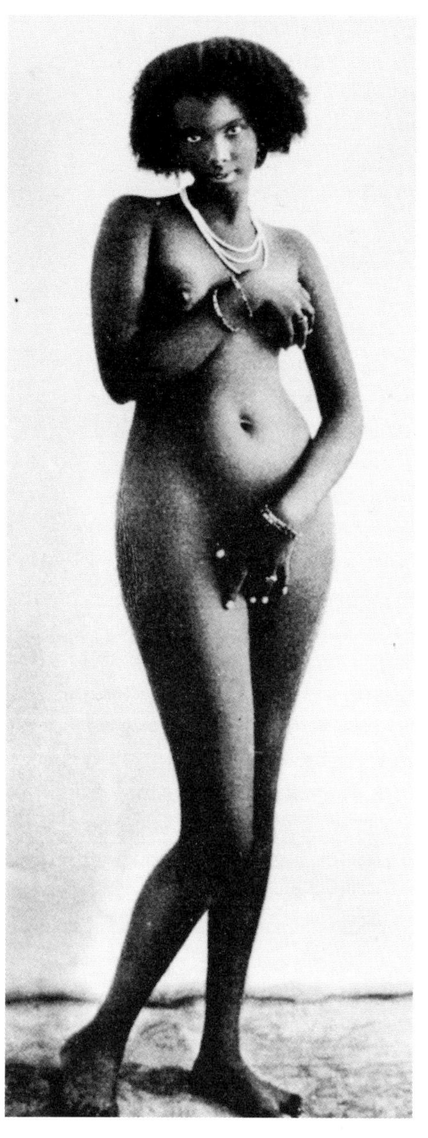

13. Photography in the service of the invader: an Ethiopian nude, depicted on an Italian postcard, produced to popularise Mussolini's invasion of 1935-6 among the troops.

Fascist Italy's impending invasion led in 1935 to a new burst of international interest in Ethiopia, which was then visited by an unprecedented number of foreign journalists and photographers.

One of the best cameramen of this period was a German, Alfred Eisenstaedt, who worked for the *Berliner Illustrierte Zeitung*. He took no fewer than 3,500 photographs in 1935, before emigrating to the United States, where he joined *Life* magazine, but returned in the following year to continue his Ethiopian photographic Odyssey.

Another photographic journalist was Ladislas Farago, a Hungarian in Britain, who reached Addis Ababa shortly prior to the opening of hostilities. His reportage was not without its difficulties. While taking a photograph in Addis Ababa shortly after his arrival he was "surrounded in a flash by a furious crowd", whereupon a policeman declared that photography was forbidden, and tried to confiscate his camera. Farago was, however, rescued by a kind, apparently European educated, Ethiopian in a car, who, then rushed off with him at top speed. The Hungarian later took many topical photographs which found their way into his *Abyssinia on the Eve* (London, 1935). It contained portraits of the Emperor and Empress; their sons, Makonnen and Sahla Sellasé; Abuna Qerillos, the head of the church; Dajazmach Balcha, a veteran of the Adwa war; the Foreign Minister, Blatténgéta Heruy; the aged Minister of War, Ras Mulugéta; the head of the Imperial Bodyguard, Balambaras Mekureya; the military commander in Harar, Ras Nasibu; the Emperor's Father Confessor, Abba Hanna; the monarch's Swedish military adviser, General Virgin; and the Italian Minister in Addis Ababa, Count Vinci. Other photographs showed Haile Sellassie driving through an Addis Ababa crowd, his palace, a group of Ministers listening to him addressing Parliament, members of the Imperial Bodyguard, one no less than 2.18 metres tall, units of the Ethiopian army, a hospital patient paying in bullets for treatment, school-children playing football, a swarm of locusts, and the storing of grain in preparation for the invasion.

An even better illustrated work was Wilhelm Goldmann's *Das ist Abessinien* (Leipzig, 1935), which featured many aspects of Ethiopian modernisation, drawn from earlier published photographs. Other pictures appeared in Anton Ziscka's *Abessinien* (Leipzig, 1935), and Ludwig Huyn and Josef Kalmer's *Abessinien Afrikas Unruhe-Herd* (Salzburg, 1935). One of the last books on pre-war Ethiopia, it contains a photograph, significantly, of Emperor Haile Sellassie practising to use a machine-gun.

One other photographically well illustrated work, produced at the end of our period, deserves mention.

Compiled by a local Greek, and printed in Egypt, Adrien Zervos's *L'empire d'éthiopie. Le mirroir de l'éthiopie moderne* 1906-1936 (Alexandria, 1936), contained over 300 photographs, of prominent personalities, public buildings, and provincial capitals, as well as historical sites, cultural activities, and foreign diplomatic missions and their staff.

Italian photographic offerings of this time were significantly different from those published elsewhere. The Fascist Government, which had decided to invade, sought to portray Ethiopia as a land of barbarism and slavery. Italian publications, which were subject to strict censorship, therefore tended to avoid reference to Ethiopia's modernisation, but nevertheless reproduced other interesting photographs. Arnaldo Cipolla showed pictures of Emperor Téwodros's cannon at Maqdala and of the Addis Ababa equestrian statue of Menilek in *L'Abissinia in armi* (Florence, 1935), while Ugo Caimpenta concentrated on "ethnic types" and foreign missionary work in *L'impero abissino* (Milan, 1935). Giannino Marescalchi's *Eritrea* (Milan 1935) concentrated mainly on the colony's "natives" and their colourful dances, as well as on Italian agricultural activity.

14. Portrait of Bedros Boyadjian, an early 20th century Armenian photographer in Addis Ababa, who took many of the photographs reproduced in these pages.

15. Photograph, taken in the 1920s by the French travellers Azais and Chambard, to illustrate an anthropological view of Amhara "racial types".

Italian photographers in this period also took many erotic pictures of Ethiopian women, designed to appeal to Mussolini's soldiers about to embark on the invasion. A series was produced as postcards by A. Baratti in Asmara, and later reproduced in Adolfo Mignemi's *Imagine coordinata per un impero. Etiopia 1935-6* (Turin, 1983). Such photographs had, however, only ephemeral circulation, for they were withdrawn in the summer of 1936 when the fascists instituted a policy of strict racial discrimination.

Many photographs reproduced in this volume were made with the approval of the persons depicted, and reflect the traditional hierarchy, but others were conceived entirely by the photographer. Those in this latter category include pictures taken by physical anthropologists, in the course of their recording of skull and other measurements. Photographs of this type are found in many publications of our period, among them Paulitschke's *Beiträge* of 1888, and Alberto Pollera's *I Baria e i Cunama* of 1913. Several anthropologists also took many photographs of considerable ethnographic interest.

Though foreign photographer-travellers played a valuable role in recording scenes of the past, they made little impact in the country itself. Even the urban population remained largely unaware of the camera until the advent, in the early twentieth century, of local photographic studios, almost entirely foreign-run.[5]

The first photographer to establish himself in the Ethiopian capital was an Armenian, Bedros Boyadjian, who arrived with an Armenian priest from Jerusalem in 1905. He was succeeded in 1928 by his son Hayjaz, who took court portraits of Empress Zawditu, and later by another son, Tony, who served as court photographer to Haile Séllassie. The second important Addis Ababa photographer was Levon Yazedjian, an Armenian from Turkey, who began work in 1909, and is accused, as we have seen, of playing an interesting, but entirely unproved, role, in Lej Iyasu's overthrow. The next photographer to come was an Indian from Gujarat, G. Mody, who began work in Harar early in the century, and travelled to Addis Ababa in 1910 where he founded a studio in the main commercial street, Ras Makonnen Avenue. A man of courage he acquired a reputation for his devotion to the sick during the Spanish influenza epidemic of 1918, and continued his photographic activities for the next three decades.

Another photographer of this period was Alex Dorflinger, a member of an Austro-Hungarian mission which arrived shortly prior to World War I, and was stranded on account of the opening of hostilities. A French resident, M. Stévénin, seeing his predicament, provided him with free lodgings, and later with the photographic materials required to establish a studio.[6] Another Armenian, Megherditch Reissian, opened a shop at about the same time in Dire Dawa, the first in the Ethiopian provinces.

Photographic activity expanded significantly in the decade or so prior to the invasion, when several foreigners, largely Armenians, established studios, mainly in Ras Makonnen Avenue, where they remained until the Italian occupation. An Armenian from Egypt, Abram Chahbaz, and a Greek, Alex Tsouklas, began work in 1923, an Armenian from Beirut, Fusan Abouseifian, in 1927, and his compatriot Hrant Vararanian at about the same time. Several photographers, likewise chiefly Armenian, also began taking cheap "minute photographs" in the market area, and thereby brought the camera for the first time within the reach of the many.

Considerable expansion of photography also took place in the Eritrean capital, Asmara, where a few Italians founded studios immediately before or after the fascist invasion. Walter Amadio, a photographer and cinema operative, set up a business in 1934, and Ambrogio Lusci became a photographer to the Italian forces shortly afterwards.

The advent of studios, and "minute photographers", introduced increasing sections of the Ethiopian public to photography. Though often conservative, more and more people began to accept the camera, and have their photograph taken, particularly at weddings and other formal occasions. The custom of carrying photographs of the deceased in funeral processions became more and more common in urban funeral processions. Increasingly widespread awareness of photography is likewise evident from the fact that the old Amharic word *se'el*, or "drawing", was given the secondary meaning of "photograph" in Joseph Baetman's Amharic-French dictionary of 1929.

The camera, in the three-quarters of a century covered by this volume, thus made immense strides. Photography had captured many remarkable images of a unique culture, and of a stunningly beautiful country and people. Many noteworthy personalities nevertheless lived and died without being photographed, and not a few Ethiopians of the present generation have no pictures of their parents, let alone of earlier ancestors.

1 C. Keller, *Alfred Ilg. Sein Leben und sein Wirken als schweizerischer Kulturbote in Abessinien* (Frauenfeld and Leipzig, 1918).

2 See also R. Pankhurst, "The Genesis of Photography in Ethiopia and the Horn of Africa", *The British Journal of Photography* (1976), Nos. 41-4.

3 On Italian photography at Massawa, and more generally Eritrea, see A. Triulzi, *L'Africa dall'immaginario alle immagini* (Turin, 1989), which contains a catalogue, by Silvana Palma, of photographic holdings at the Turin Biblioteca Reale. See also S. Palma, "La fototeca dell'Istituto Italo-Africano: appunti e problemi di un lavoro di riordino", *Africa* (1989), XLIV, 595-609, and Triulzi, "Preliminary Report on Two Photographic Collections in Italy", in A. Roberts, Photographs as Sources for *African History* (London, 1988).

4 See also R. Pankhurst, "The Political Image: The Impact of the Camera in an Ancient Independent African State", in E. Edwards, *Anthropology and Photography* 1860-1920 (New Haven, Conn. and London, 1992). On Iyasu's overthrow see Gebre-Igziabiher Elyas and R. K. Molvaer, *Prowess, Piety and Politics. The Chronicle of Abeto Iyasu and Empress Zewditu of Ethiopia* (1909-1930) (Köln, 1994), pp. 354-5. Dr Bairu Tafla, one of many scholars with whom we have discussed the question of Iyasu's photographs, ventures the hypothesis that the prince, not above a juvenile prank, enjoyed being photographed in various guises, and, unlike later commentators, did not attach any religious significance to the matter.

5 On the development of photographic studios see A. Zervos, *L'empire d'Ethiopie* (Alexandria, 1936).

6 On this photographer see D. Pariset, *Al Tempo di Menelik* (Rome, 1937).

I. Historic Personalities

THE COMING TO ETHIOPIA of the camera in the second half of the nineteenth century led to the photographing of two of the country's three great nineteenth century modernising emperors. The reforming Emperor Téwodros II, who reigned from 1855 until his dramatic suicide at Maqdala in 1868, is known only from drawings and engravings. Several photographs were, however, taken of his successor Emperor Yohannes IV, whose reign began in 1872, and came to an end with his death at the battle of Matamma in 1889. There are by contrast innumerable photographs of Menilek II, a particularly photogenic figure, who ruled as King of Shawa from 1865 until his accession to the imperial throne in 1889, and as Emperor up to his death, after long years of paralysis, in 1913.

A few other personalities were photographed at the beginning of the period. They included the Oromo, or Galla, Queen Mastewoat of Wallo, with her son, and Téwodros's orphaned son Alamayahu. His portrait was taken first in Ethiopia, and later in Britain, where he arrived with the returning Maqdala Expedition in 1868. He died while still a young man without ever returning home. A photograph was also later taken of Emperor Yohannes's son Ras Araya Sellasé, who predeceased his father without having played a major role in state affairs.

There are on the other hand numerous photographs of virtually all the principal personalities of Menilek's reign. They include his powerful, and staunchly patriotic, consort Empress Taytu, who played a major role in state politics. Portraits also exist for many prominent governors and other rulers, several of whom appear as figures larger than life. Personalities photographed include a wide variety of chiefs from virtually all parts of the far-flung empire: Ras Makonnen, governor of Harar (and father of the future Emperor Haile Sellassie), Ras Mangasha, of Tegray, and Ras Mika'él, of Wallo, besides King Takla Haymanot of Gojjam, King Tona of Walamo, also known as Walayta, King Gaki Sherocho of Kafa, Abba Jifar of Jimma, Shaykh Khojoli of Bela Shangul, and King Wuria Maru of Gambela. Portraits were also taken of many other notables of the time, among them Ras Alula, hero of the battles of Dogali in 1887 and Adwa of 1896; Abuna Matéwos, the Egyptian Coptic head of the Ethiopian Orthodox Church; foreign educated Ethiopians, such as Nagadras Gabra Haywat Baykadagn and Tassama Eshaté, not to mention such leading foreigners as the Swiss engineer-cum-diplomat Alfred Ilg, the French merchant and railway promoter Léon Chefneux, and the Armenian entrepreneur Sarkis Terzian.

Numerous photographs were later taken of Menilek's grandson and heir, Lej Iyasu, an enigmatic ruler who was never crowned. A particularly handsome, and photogenic, but politically unsuccessful young man, he was the subject of many interesting photographs: with his father Ras Mika'él, ruler of Wallo, with his guardian Menilek's trusted courtier Ras Tassama, and with his rival and later successor Ras Tafari Makonnen, the future Emperor Haile Sellassie. Iyasu, a young man of many parts, is likewise seen dressed as a Christian warrior and nobleman, as a potentate surrounded by Harari Muslim dignitaries, and scantily clothed as an Afar, or Dankali, tribesman. Iyasu in the course of a short reign incurred the antagonism of the Church, the Shawan nobility, and the Allied Powers during World War I. He was deposed in 1916.

After Iyasu's overthrow power was shared between Menilek's daughter Empress Zawditu, who came to the throne in 1916, and Ras Tafari, who was nominated Heir to the Throne, and succeeded her as Emperor Haile Sellassie I in 1930. Both rulers, the conservative Empress, and the modernising Tafari, sought in their own way to continue Menilek's state-building policies, and were the subject of numerous portraits. Many pictures were also taken of the Emperor's impressive coronation in 1930, of his wife, Empress Manan, and family, besides courtiers, dignitaries, ministers, and provincial rulers. The latter, like those in Menilek's service, came from all over the country, and included Ras Seyoum of Tegray, Ras Haylu of Gojjam, Dajazmach Habta Maryam of Wallaga, and the Afar leader Dajazmach Yaho Muhammad. Photographs are likewise extant for such notable figures as Ras Kasa and Ras Emeru, the Coptic prelate Abuna Qerillos, the French-educated artist Agagnaw Engeda, and Hakim Warqnah, also known as Dr Martin, who served as his country's ambassador in Britain during the Italian fascist invasion of 1935-6.

16. Maqdala, Emperor Téwodros's mountain fortress and capital in 1868: an early picture taken by the photographers attached to the British expedition.

17. Emperor Téwodros's mortar, cast in Ethiopia, at Maqdala: an early 20th century photograph.

18. Houses at the top of Maqdala: another photograph by the British expeditionary force.

19. Emperor Téwodros's European craftsmen, cannon-builders, and sometime hostages, after their release from Maqdala, April 1868.

20. General Robert Napier, seated centre, with officers commanding the Maqdala expedition of 1867-8.

21. Spick and span at the beginning of the Maqdala expedition: British troops and their tents.

22. Marcha Warqé, envoy of Dajazmach Kasa (the future Emperor Yohannes IV), centre, photographed with his followers, during his talks with the British expedition in 1867. With him is British Brigadier-General Merewether. Beside the latter, his Swiss interpreter Munzinger and, seated left, his adviser Captain Speedy.

23. The church at Maqdala where Emperor Téwodros was buried. Two British soldiers on guard.

24. An Ethiopian minstrel playing the masenqo, or one-stringed fiddle: an early photograph taken by the British expedition.

25. Mastawat, the Oromo queen of Wallo, with her eldest son, Imam Ahmad, and, standing, a leading courtier.

26. Emperor Yohannes IV, with his son Ras Araya Sellasé. Note the Emperor's royal crown.

27. Ras Alula Abba Naga, victor of the battle of Dogali, 1887, and one of the generals at the battle of Adwa, 1896.

EMPEROR MENILEK (KING OF SHAWA, 1855–1889, EMPEROR OF ALL ETHIOPIA, 1889–1913) AND HIS TIMES

28. Menilek, the then young king of Shawa, with his warriors in military dress.

29. Menilek's consort Queen Taytu Betul, as a young woman.

30. Emperor Menilek in coronation garb.

31. Empress Taytu in regal attire. Note, her imperial crown.

32. Emperor Menilek, by then an old man, with Queen Taytu Betul. Note, his characteristic headgear.

33. Menilek, travelling by mule. Note umbrella, symbol of royalty.

34. The aged Menilek, photographed holding a staff, with his dog. A little known picture taken in December 1911 by James William Young, of Aden.

35. Taytu with apparently the same dog.

36. Menilek looking down from the church at Addis Alam.

37. Menilek and his followers.

38. Menilek, with his courtiers, in ceremonial dress. They include Dajazmach Balcha Safa (4th from left), Ras Tasamma Nadaw (6th from left), Ras Walda Giyorgis (7th from left), and Ras Abata Bwayalaw (4th from right).

39. Ras Makonnen, with his followers, in London, for the coronation of King Edward VII in 1901. A studio photograph by Lafayette.

40. Abuna Matéwos, Patriarch of the Ethiopian Church.

41. Ras Mika'él, formerly Muhammad Ali, of Wallo, and father of Lej Iyasu.

42. Ras Mangasha Yohannes, son of Emperor Yohannes IV, and ruler of Tegray.

43. Ras Makonnen Walda Mika'él, governor of Harar, and father of Emperor Haile Sellassie.

44. King Takla Haymanot of Gojjam (reigned 1881-1901).

45. Fitawrari Habta Giyorgis Dinaglé, Menilek's Minister of War.

46. Nagadras Hayla Giyorgis Walda Mika'él, Menilek's Minister of Foreign Affairs and Trade.

47. One of Menilek's principal generals, Ras Tasamma Nadaw.

48. Menilek's court chronicler Sahafé Tezaz Gabra Sellasé Walda Aragay.

49. Dajazmach Balcha Safo, a hero of the battle of Adwa.

50. Another of Menilek's generals, Ras Walda Giyorgis, photographed with his favourite rifle.

51. Menilek's Swiss adviser, Alfred Ilg, welcoming the German diplomatic mission, 1905.

52. Gaki Sheroko, King, or Tati, of Kafa.

53. Tona, King, or Kawa, of Walayta, then better known as Wolamo.

54. Shayk Khojali al-Hasan, ruler of Bela Shangul, also known as Beni Shangul.

55. Abba Jifar II, ruler of Jimma, with his wife Genne Mingitti.

56. Wuria Maru, the local ruler of Gambéla, with his family.

57. Dajazmach Gabra Sellasé Bareya Gaber, of Tegray with an envoy from the Italian colony of Eritrea.

58. Kantiba Gabru Dasta, who was educated in Switzerland and Germany, with his eldest daughter.

59. Nagadras Gabra Heywat Baykadagn, a young intellectual of the latter part of Menilek's reign.

60. Tassama Eshaté, a German-educated sculptor, poet and amateur photographer.

61. Seven-year-old Lej Iyasu reading the Psalms of David, with other young students at Ankobar.

62. Photograph with a political message: Young Lej Iyasu, Menilek's appointed heir, with his guardian Ras Tassama Nadaw. The latter's presence in the picture serves to show that Ethiopia's destiny will be in tried, as well as youthful hands.

63. Iyasu, a studio portrait.

64. Iyasu shows his filial devotion to his father Ras Mika'él of Wallo. Note Menilek's crown on right.

65. Another photo with a political message: Iyasu with his grandfather, Emperor Menilek, who had designated him as his heir.

66. Iyasu with his niece Manan Liban, wife of Tafari Makonnen, the future Emperor Haile Sellassie.

67. Lej Iyasu with Dajazmach Tafari, the future Emperor Haile Sellassie. Note Iyasu, shoeless, in royal pose, with sword; Tafari, with shoes, his clothing arranged to show his deferential status.

68. Another politcally interesting photo:
Iyasu, wearing a Muslim turban, photographed at Harar, in the house of Abdulahi Ali Sadiq, a prominent man of the city, standing right.

69. Iyasu in Afar, or Danakil, dress:
a photograph taken during one of the prince's travels in Ethiopia's eastern lowlands.

70. Lej Iyasu, with Dajazmach Tafari Makonnen, left, and Dajazmach Berru Wald Gabr'èl, right.

71. Iyasu in regal dress. Note, grand cordon, with cross aound his neck.

72. Empress Zawditu Menilek, with her courtiers and attendants.

73. Another photograph of Zawditu with attendants.

74. Zawditu, with one of her most favoured priests.

75. Zawditu in royal robes.

76. Crowds at Entoto waiting to accompany Zawditu back to her palace at Addis Ababa.

EMPEROR HAILE SELLASIE (CROWNED 1930) AND HIS TIMES, UP TO THE 1935–6 INVASION

77. Young Tafari Makonnen, the future Emperor Haile Sellassie, with his father, Ras Makonnen Walda Mika'él,

78. Young Tafari dressed as a church school student.

79. Tafari, with his cousin Emru Haile Sellassie, the future Ras, extreme right, in their youth.

80. Tafari, aged 15, with two courtiers, Dajazmach Haile Sellassie and Abba Tabor.

81. Tafari, the Heir to the Throne, left, and Zawditu, the actual sovereign, right, hold court several paces apart.

82. Ras Tafari's European tour of 1924: at the palace of Versailles, under a bas relief of King Louis XIV.

83. Ras Tafari at the League of Nations, which Ethiopia had succeeded in joining in the previous year.

84. Inspecting an aeroplane at Le Bourget airport, France.

85. Negus Tafari, just about to become Emperor Haile Sellassie, welcomes Marshal Franchet d'Espery, who has come to Addis Ababa for his host's coronation as emperor in 1930.

86. Arrival of Britain's Duke of Gloucester for the coronation.

87. Coronation ceremonies inside the palace: Emperor and Empress, each under a separate dias.

88. Tafari, as Negus, or King.

89. Studio portrait of the Imperial Family. Crown Prince Asfa Wassan stands between his parents, with younger brother Makonnen on his father's left.

90. Haile Sellassie, in royal dress, with crown and orb.

91. The Emperor crowned.

92. Coronation ceremonies in Addis Ababa, 1930. Note temporary triumphal arch.

93. Coronation parade, Addis Ababa, 1930.

94. Emperor Haile Sellassie and Empress Manan at their Addis Ababa palace in 1935. Note royal lion.

95. On the eve of the Italian fascist invasion the Emperor, symbolically dispensing with a rifle-carrier, carries his own weapon as he walks ceremonially around St George's Cathedral, Addis Ababa.

96. The Emperor, in military uniform, as commander of the Ethiopian army. October 1935.

97. Ras Gugsa Walé, of Bagémder.

98. Ras Kasa Haylu, of Shawa.

99. Ras Haylu Takla Haymanot, of Gojjam.

100. Ras Seyum Mangasha, of Tegray.

101. Ras Mulugéta Yeggazu, Minister of War.

102. Ras Emeru Haile Sellassie, with his family.

103. Abuna Qerillos, Patriarch of the Ethiopian Church.

104. Dajazmach Habta Maryam, of Wallaga.

105. Fitawrari Yaho Muhammad, of Afar.

106. The French-educated artist Agagnaw Engeda.

107. Priests prostrating themselves in honour of Abuna Qerillos.

108. Haile Sellassie visiting the Indian commercial house of Muhammad Ali, the merchant seen standing on the monarch's right.

II. Historic Towns: North, South, East and West

ETHIOPIA, a country virtually as large as France and Spain combined, has traditionally been a land of villages and isolated homesteads. Over the centuries there nevertheless emerged a number of important towns and cities, for the most part situated in the country's cool temperate highlands. These settlements, an important facet of the country's rich cultural heritage, were the subject of numerous photographs dating from the last years of the nineteenth century or the first decades of the twentieth.

The earliest known important Ethiopian settlement was Yéha, in the province of Tegray, the country's northernmost region. Yéha, which dates back to the fifth century BC, or even earlier, is the site of an imposing pre-Christian temple, doubtless erected in honour of pagan gods, also worshipped in South Arabia.

The great city of Aksum, also situated in Tegray, must be ranked as Ethiopia's first historic capital. It probably came into existence a century or so prior to the birth of Christ, and was the capital of the Aksumite empire, an internationally important polity which traded as far as Egypt and India, and issued its own currency. The site, it is believed, of the country's first major church, erected early in the fourth century, Aksum was long a major religious centre, and was described by Theodore Bent, a British nineteenth century traveller, as the Holy City of the Ethiopians. Its antiquities include famous stone obelisks, one of which was looted by the Italian fascist dictator Mussolini in 1937 and taken to Rome, multi-lingual stone inscriptions dating back to the early fourth century, and excellently fashioned underground tombs. Also of considerable interest is the famous Church of St Mary of Seyon, which was rebuilt in castellated style in the sixteenth and seventeenth centuries. Ethiopian tradition claims that it is to this day the repository of the Ark of the Covenant.

Our photographic tour then proceeds southwards to other parts of Tegray, with scenes of the city of Adwa, which rose to prominence in the eighteenth century, and of the town of Maqalé, site of a well-preserved palace of the Emperor Yohannes. Several photographs of Tegray housing have been included to show the diversity of architecture found within even one of Ethiopia's regions.

Attention then shifts southwards to Lalibala, site of eleven of Ethiopia's most famous rock-hewn churches. Believed to have been erected by the thirteenth century monarch of that name, these structures were first described by the early sixteenth century Portuguese traveller Francisco Alvares, who declined to write much about them lest he not be believed. We likewise restrict our photographic presentation of Lalibala - in the hope that readers will visit it, and see for themselves these churches which are considered among the wonders of the world.

Gondar, our next important photographic "port of call", in the north-west of the country, was established as the Ethiopian capital by Emperor Fasiladas in the early seventeenth century. He it was who erected the first of many grand castle-like palaces, some of the most remarkable in sub-Saharan Africa, which to this day adorn the city. Gondar, which lay on the trade routes to the Sudan and the port of Massawa, came into existence after several centuries of "moving capitals". It remained the Ethiopian metropolis - and a major commercial, religious and cultural centre, with many remarkably fine churches - until the late nineteenth century reign of Emperor Téwodros.

Of no less interest is the old Islamic walled city of Harar, situated in the south-east of the country. Erected on the trade route to the Gulf of Aden ports, the city became a great commercial emporium, issuing its own currency, and was renowned for both its fine book-binding and its beautiful baskets. Harar, a centre of Muslim learning, was long protected by its stout walls, erected in the mid-sixteenth century, which were pierced by five gates. The city was surrounded by well-cultivated gardens, growing coffee, a noted speciality of the area, and the mild narcotic chat.

Other localities depicted in this chapter include Dabra Tabor, which Emperors Téwodros and Yohannes both for a time used as their capital; Ankobar, site of the principal nineteenth century capital of Shawa; Dasé, the historic capital of Wallo; Jimma, one of the most important political and commercial centres of the west; and the railway town of Dire Dawa, established in 1902.

109. Yéha: The ancient Sabaean temple, dating back to perhaps the 7th century BC.

110. Aksum, an early 20th century procession of priests.

111. The church of St Mary of Seyon, or Zion, Aksum.

112. The city of Aksum in 1906. Note the church of St Mary, centre, and predominance of thatched houses.

113. The largest standing obelisk at Aksum.

114. Aksum: thatched houses around the standing stele.

115. The market town of Adwa in 1935.
Note the double concentric walls around the church of Madhané Alam, or Saviour of the World.

116. The church of Madhané Alam, Adwa.

117. Maqalé: the palace of Emperor Yohannes IV surrounded by two concentric walls. A late nineteenth century photograph.

118. Emperor Yohannes IV's palace at Maqalé.

119. Lalibala: the church of Madhané Alam, or Saviour of the World.

120–123. Houses in early twentieth century Tegray. Note external stairs and second storeys.

124. The great 17th century castle of Emperor Fasiladas at Gondar.

125. Gondar: the northern side of the palace complex photographed in 1905.

126. Gondar: the early 18th century palace of Emperor Bakaffa.

127. Dabra Tabor, the old capital of Bagémder, with its church on the hill-top.

128. Dasé, capital of Wallo, with the palace in the distance.

129. Ankobar, the 18th and early 19th century capital of Shawa: the palace hill.

130. The old walled city of Harar, a town of minarets, in 1935.

131. Harar: the newly established church of Madhané Alam in 1897.

132. The mosque and minarets of Harar in 1935.

133–4. Two of Harar's historic gates as seen in 1901.

135. Harar: Ras Makonnen's palace in 1903.

136. The railway town of Dire Dawa: glimpse of the mosque in the 1920s.

138. Jimma: the sultan's palace.

137. Dire Dawa: the commercial quarter.

III. ADDIS ABABA: THE "NEW FLOWER"

ETHIOPIA'S PRESENT-DAY CAPITAL, Addis Ababa, is by Ethiopian standards a very recent city. It came into existence only a little over a century ago.

The origins of Addis Ababa date back to the early 1880s when Menilek, then ruler of Shawa province, moved his capital from Ankobar, a couple of day's march to the north, and established a camp at Entoto, in the mountains above today's Ethiopian metropolis. The site was in part selected because it had been the capital of the early sixteenth century Emperor Lebna Dengel, as evident from a rock church and other antiquities in the area.

Entoto was strategically well situated, but unsatisfactory as a permanent capital: its mountainous location made it difficult of access, and caused it to suffer during the rainy season from severe tropical storms. The Finfine plain to the south was by contrast in every way attractive. It had an equitable climate, and fertile, well-watered, land. It was an area moreover where warm thermal water gushed out of the ground, at Felweha, literally "Boiling Water". This water, which can today be savoured at the Ghion, Hilton and Sheraton hotels, attracted Menilek's courtiers, especially to his consort, Queen Taytu, who loved "taking the waters" there.

The Filweha area was at first the site mainly of tents, but Taytu soon afterwards erected a palace there, and the courtiers followed suite. A new settlement thus emerged, which Taytu named Addis Ababa, literally "New Flower".

Addis Ababa, situated on the lower slopes of the Entoto mountain range, and bisected by streams and gorges running from north to south, was originally a scattered settlement, with two main nuclei, both illustrated in these pages.

The first was Menilek's great palace, situated on the top of a hill on the settlement's eastern side. The royal establishment consisted of a group of buildings in a compound measuring some two kilometres by one and a half. This large space contained the residence of the Emperor and Empress, a great banqueting hall, storehouses and buildings for the preparation of food and drink, besides administrative and other offices. The palace, always crowded and a place of much bustle, was the site of all the political, diplomatic, ceremonial and other activity associated with the government of an empire. This included huge banquets, at which as many twenty thousand guests would be fed in a single day, in several sittings.

Menilek's personal servants and attendants, and those of his consort, lived in much humbler dwellings in the vicinity of the palace. The residences of the principal chiefs were located on neighbouring hilltops, with their servants and other followers clustered in huts around them.

The settlement's other centre, to the west, below the Church of Saint George, was the market, one of the largest in Africa. It handled an infinite variety of goods, including cattle, grain, and other provisions from the surrounding countryside; bars of rock salt, which served instead of money, from the Afar depression in the far north-east; gold, ivory, civet musk, and in the early days slaves, from the lands of the south-west; and manufactured articles of all kinds, among them textiles and firearms, imported from Europe, the United States, and India.

The Addis Ababa market, like those in other parts of the country, was at first open to the sun and rain. Covered stalls were, however, later erected - and many foreign merchants, Indian, Arab and Armenian, erected sizeable, in some instances quite impressive, dwellings and storehouses.

Addis Ababa in the period covered by our photographs underwent many major changes. What had begun as a camp of tents witnessed the erection of more and more permanent structures, the finest made of stone with wooden balconies and balustrades. The settlement, which consisted at first of dispersed encampments, with herds of cattle, horses, mules and other livestock in the spaces between them, gradually merged into a more integrated whole. An essentially treeless area was transformed, by the introduction of the fast-growing Australian eucalyptus tree, into a largely wooded town, which Dr Mérab, a Georgian resident, described around the time of World War I as a Eucalyptopolis, or Town of Eucalyptus Trees.

The city in the early twentieth century was the site of many innovations: an elegant railway station and the solid headquarters of the national bank, as well as a post office, a taxi-rank, a fine hotel founded by Empress Taytu, several cinemas - and a number of foreign legations. Ethiopian men and women were, however, still in their traditional dress.

139. Entoto, precursor of Addis Ababa: the church of Maryam, or St Mary, in the early 20th century.

140. The church of Ragu'él at Entoto, at the end of the 19th century.

141. Entoto: a nobleman's house.

142. Felwaha, the southern quarter of Addis Ababa in 1906-7. Note, the thermal springs, foreground, which led to the founding of the city, as well as the absence of houses and trees.

143. The hot springs of Felwaha.

144. The palace and surrounding buildings.

145. The palace clock tower, with Menilek's artillery, some of it captured at the battle of Adwa in 1896.

146. Another view of the palace.

147. The palace and surrounding buildings.

148. Foreign diplomats leaving the palace after an interview with Menilek.

149. The palace's huge three-gabled banqueting hall.

150. One of Menilek's great banquets. Note photographs of the Emperor and Empress, on poles to the left and right.

151. St George's church, the city's principal religious establishment, situated in the market area, in the early twentieth century.

152. Ras Makonnen bridge, linking the palace with the commercial centre. Note thatched houses and the then virtual absence of buildings.

153. The market area at the turn of the century: beating the drum to announce a royal decree. Note national flags consisting of three separate pennons.

154. An Indian house in the market area.

155. An Indian store in the same area.

156. Part of the city's commercial quarter in the early 1930s.

157. The railway station, showing the then recently erected statue of the Lion of Judah, 1930.

158. Addis Ababa, as seen from the railway station in 1930. Note Lion of Judah, centre, and behind it a track (the future Churchill Road) leading to the commercial centre. Note also eucalyptus trees, and general absence of buildings.

159. The city's first Post and Telegraph Office.

160. The city's second Post and Telegraph Office. Note the presence by then of motor cars.

161. The Bank of Abyssinia's first premises, established 1905.

162. The bank's second, and more solid, building, erected 1910.

163. The Etegé Hotel, founded by Empress Taytu in 1907.

164. The British Legation, housed in traditional-style local dwellings, erected soon after the opening of Anglo-Ethiopian relations in 1897.

165. The Ethiopian capital around 1920, still a rural town.

166. Aerial view of Addis Ababa: the palace area in 1930. Note eucalyptus trees.

167. The city in 1930, with the railway station, centre, the palace compound, top-right, and the market area, top-left. Note open spaces on either side of the future Churchill Road and in the foreground.

IV. Economic, Social and Cultural Life: Tradition and Diversity

Ethiopia, in the period covered by this volume, was an almost entirely agricultural and pastoral country, the beauty, and diversity, of which can only inadequately be suggested in these pages. Agriculture in the northern and central highlands was based largely on ox-drawn ploughs, though various types of "digging sticks" were also in use in some areas. Harvesting was usually carried out with the help of sickles. Crops would be protected from monkeys and other animal predators by youths with slings, while threshing would be carried out by cattle who would walk round and round on rural threshing floors.

The country had extensive livestock. Many cattle were of the zebu, or hump-backed, variety, also common in India. Some had huge horns, often used as drinking vessels for beer.

Despite the predominance of agriculture, craftwork was also important. Artisan included blacksmiths, weavers, and potters, the last-mentioned mostly women. Handicraft work in many areas was carried out by members of religious or other minority groups, among them Muslims and Judaic Ethiopians, known as Falashas, or Béta Esra'él, most of whom have since emigrated to Israel. House-building was on the other hand generally carried out by the agriculturalists themselves, who often gathered together to undertake the work communally.

Many parts of the country were renowned for their artefacts. These included pottery and basketry, as well as beautiful horn, silver, and, for the very privileged, gold, jewellery. Gold was for the most part locally panned in rivers of the west, while silver came from melted Maria Theresa thalers imported from Austria. Ivory was obtained from elephants, whose numbers, on account of the depredations of hunters, were fast diminishing.

Trade was mainly carried out in markets which were found all over the country. Most were held on a weekly basis, which enabled itinerant traders to travel frequently from one to another and do business throughout the year. Larger towns had sizeable daily markets, with sections specialising in different types of merchandise: salt bars, baskets, pottery, cattle, grain, etc.

The Ethiopian countryside was in many areas rugged, and difficult to traverse, the more so as there were few roads or bridges until modern times. The Ethiopians of former days nevertheless travelled extensively, both in peace and war. The peasantry were obliged in many cases to walk long distances to bring supplies to market. Firewood and water, both necessities of life, had often to be obtained far away from homesteads, whence women had to carry them on their backs. Trade was largely in the hands of itinerant merchants, who travelled with their goods from one region to another. Merchandise in the highlands was transported mainly by mules and donkeys; in the lowlands, by camels. Supplies were also often carried by porters, especially slaves. Crossings of rivers, and lakes, presented many difficulties, which, as our photographs show, were overcome by various means.

Ethiopian men and women, depending on their age, region, ethnic background, and occupation, displayed a considerable variety in their style of dress, and plaited their hair in many different ways.

The Church throughout this period played a major role in social and cultural, as well as in religious life. The inhabitants of the Christian highlands included a large population of priests, dabtaras, or lay clerics, and deacons, as well as monks, nuns, and hermits. Lay members of the community were also deeply influenced by Christian values. They were baptised and circumcised in accordance with Biblical tradition, celebrated all the Christian festivals, rigidly followed the fasts stipulated by the Church, and were buried whenever possible in consecrated ground. Education in the Christian parts of the country was carried out in church schools, and in Muslim areas in Qoranic schools. Such religious institutions until modern times constituted the sole sources of education. Many Christian youngsters opted to become wandering students, and, dressed in simple sheep's cloaks, travelled widely in search of learning.

Other aspects of traditional Ethiopian social life illustrated in this volume include the custom whereby thieves were identified by a lebashay, or thief detector, and the practice of chaining debtors and creditors together. Mention may also be made of the country's old-time musical instruments, which included string instruments, most notably the bagana, often referred to as the Harp of David, wind instruments, and drums. Photographs of two different games are also reproduced: santaraj, a form of chess, played by the nobility, and gabata, a board-game also widely known in other parts of Africa, and Asia, and often internationally referred to as mancala.

168. Ploughing in the Ethiopian highlands, in the early 20th century.

169. Use of digging sticks, in the Harar region.

170. A youngster with a sling protects the family's crops against birds, monkeys and other predators.

171. Ethiopian long-horned cattle.

172. In the Afar lowlands: a herd of camels by the thermal sping of Bilén in the late 19th century.

173. Woman cleaning grain with child on her back.

174. Women slaves grinding grain for a great Ras's household.

175. Typical women's work: two country-women carrying water-jars on their back.

176. Cutting up a bull for a feast.

177. Eating with their hands around a masob, or basket table. Note berellé, or glass bottles, containing taj, or honey wine, and pot with wat, or stew.

178. House construction.

179. Three weavers, seated in the open air, having dug holes in the ground for their feet, work their traditional looms.

180. Falasha potters, with their work, 1905.

181. Silver crosses and other jewellery.

182. Examples of variously shaped knives and daggers from different regions or ethnic groups, Danakil, Habab, Bela Shangul.

183. Addis Ababa market early in the twentieth century.

184. The market, Addis Ababa. Note women with straw parasols.

185. Addis Ababa's pottery market.

186–8.
Three views of the Addis Ababa market in the late 19th and early 20th centuries: sale of spices, salt bars, and baskets.

189. The grain market, Maqalé.

190. The meat market, Harar, 1905.

191. The Gondar market: weighing raw cotton, 1905.

192. A bare-footed peasant-soldier, in the early 20th century.

193. An early 20th century camel caravan from the coastal lowlands enters Addis Ababa.

194. Bare-breasted women porters from the south carrying firewood.

195. Two noblemen, on mules, in front of an aristocrat's house.

196. An Addis Ababa street scene.

197. A tanqwa, or traditional papyrus boat, near the shores of Lake Tana.

198. An old-style bridge, made of creepers, over the Gojab river.

199. A traditional letter-carrier stops for a quick meal and drink. Note agagal, or portable food-basket, and his letter held in a cleft stick.

200. Lepers at the Capuchin leprosarium at Harar: Brother Théotime in attendance.

201. Traditional Ethiopian bone-setting, the patient evidently in some distress.

202. A peasant woman with her child.

203. A young Addis Ababa woman, with Oromo hair-style.

204. Two noblewomen: Lej Iyasu's daughter, Alam Sahay, left, with her mother, Sabla Wangél Haylu.

205. An aristocratic woman, dressed in a finely decorated cloak.

206. A countryman, with traditional burnous, made of thickly-woven black wool. Note hood, sometimes used to protect rifle from the rain.

207. Fitawrari Kinfu Kidane, a highly respected goldsmith of Adwa.

208. An Afar elder, with warrior companions.

209. Two nobles with lions' hair head-dress and embroidered cloaks.

210. A Guragé woman, with an ornate hair-style.

211. An Oromo woman.

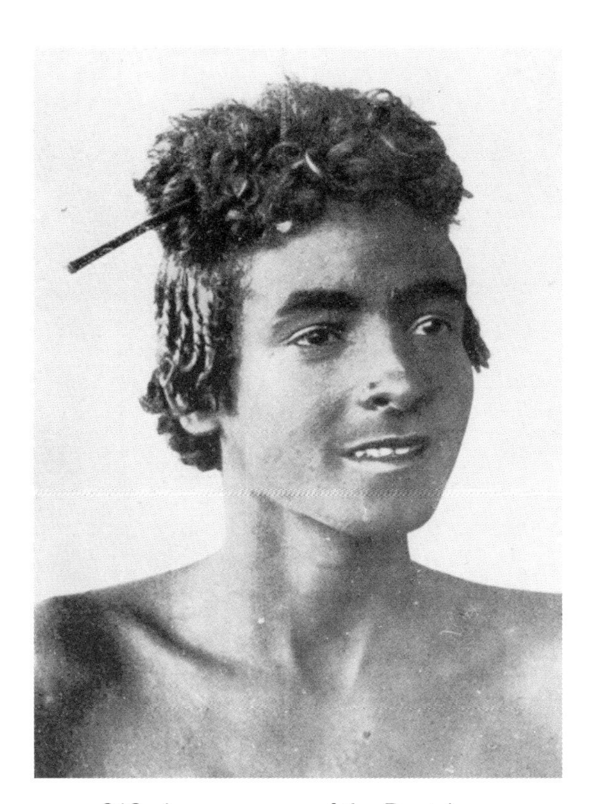

212. A young man of the Bani Amer.

213. A beaded woman from the south-west.

214. A Kafa woman's elaborate hair-style.

215. An early 19th century Addis Ababa waif. Though bare-footed he has two silver crosses and an amulet tied round his neck.

216. Four elephant-hunters.

217. Early 20th century ivory merchants making their purchases.

218. Ras Bazabeh's soldiers at Dabra Marqos, Gojjam.

219. Gojjamé troops ready to depart, early 20th century. Note flag composed of three separate pennons.

220. Four Shawan nobles mounted for departure, with their attendants.

221. A Maji chief with his followers.

222. An early 20th century veteran warrior, with lion's hair head-dress and lamd, or cloak.

223. Qanyazmach Kabala, one of Emperor Haile Sellassie's respected guards, in full splendour.

224. A Lalibala, or wandering monk.

225. A nobleman from the south-west, with cloak and fly-whisk.

226. An Afar warrior.

227. Priests of St Mary of Seyon in 1905, displaying their church finery, with crowns and dabab, or processional umbrellas.

228. An old priest, with hand-cross, prayer-book and prayer-beads.

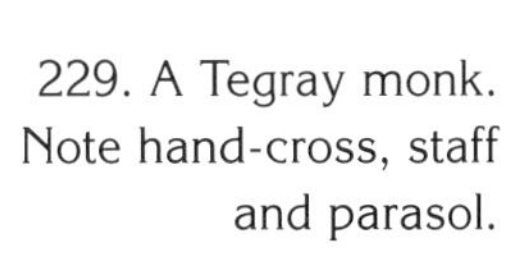

229. A Tegray monk. Note hand-cross, staff and parasol.

230. Wandering church students, at Boruméda, early 20th century.

231. An open-air church school at Washara Maryam monastery, specialising in qéné, or religious poetry.

232. A small group of pilgrims on a journey to the great religious centre of Lalibala.

233. Emperor Menilek and his followers after divine service.

234. A wandering monk.

235. Masqal, the Feast of the Cross: priests and others gather around the pole covered with branches to be burnt.

236. Easter: The stately dance of the priests, said to emulate that performed by King David in front of the Ark in Biblical times.

237. A léba shay, or traditional thief-catcher: a young boy, in front, is held on a sheath, and leads his companions to the supposed criminal.

238. The quragna system, in which plaintiff and defendant were chained together until justice was done.

239. A nobleman's court of justice.

240. A public hanging, in a tree.

241. A flogging, in the Addis Ababa's market area.

242. A musical company, with embilta, or long flutes, left, and masenqo, or one-string fiddles, right.

243. Playing the bagana, or "Harp of David".

244. A kabaro, or kettledrum, used in church services.

245. The board-game gabata, as played in Tegray. Note, folding wooden board, with three rows, each with six holes.

246. Santaraj, or traditional Ethiopian chess, as played by the nobility. An early 20th century photograph, with Dajazmach Ali, centre, and Dajazmach Gabra Sellasé, right.

V. INNOVATION AND MODERNISATION

AFTER THE ADVENT OF FIREARMS in the fifteenth century, a succession of Ethiopian rulers sought to import European weapons, and even to adopt European military techniques. Such initiatives were, however, largely unsuccessful, mainly because of the country's lack of effective access to the sea, which rendered contacts with the outside world difficult and at best intermittent.

The coming of the Industrial Revolution in eighteenth century Europe ushered in changes which were to have profound consequences for Ethiopia, as for the rest of the world. The modernisation of the country's northern neighbour Egypt, under her reforming ruler Muhammad Ali, would probably have been followed by a similar drive in Ethiopia had it not then been suffering from disunity and civil war. The land-locked and war-torn country in consequence underwent little or no technological development, and was accordingly outstripped by other lands. The strategic consequences of this shift in the balance of power was symbolised by the British seizure of Aden, and by the increasing presence of European steam-driven warships in Red Sea and Gulf of Aden waters, as well as by Egypt's conquest of the Sudan. This was followed by Egyptian annexation of border territories at Ethiopia's expense.

Emperor Téwodros, the first of the three great nineteenth century Ethiopian rulers, was well aware of his country's need of modernisation, especially in the military field. He attempted to reorganise his army, persuaded a group of European missionaries to make him artillery, and undertook some road building. His attempts at modernisation, however, soon floundered, largely as a result of his disastrous conflict with the British Government. This led, with the seeming inevitability of a Greek tragedy, to the despatch of the Napier expedition, and to his military defeat, and suicide, at Maqdala in 1868. Emperor Yohannes was likewise interested in modernisation, but his efforts in that field were frustrated by conflicts with foreign invaders at the coast: with the Egyptians in the 1870s, and the Italians in the 1880s. By the time of his death at the battle of Matamma in 1889 the country was technologically scarcely more advanced than at the beginning of the century.

Ethiopia in 1889, the year of Emperor Menilek's assumption of imperial power, was thus in desperate need of modernisation, if only to maintain her independence in the era of the European Scramble for Africa. Menilek took steps in the early 1890s to introduce Ethiopia's first national currency and postage stamps, despatched a handful of youngsters abroad for foreign studies, and granted a concession for a railway to the Gulf of Aden port of Jibuti. Most of his other innovations were introduced, however, only after his resounding victory over the Italians at Adwa in 1896, and in fact date mainly from the early twentieth century.

These innovations included the establishment of the railway, which, beginning at the coast, reached the town of Dire Dawa in 1902, and the vicinity of Addis Ababa in 1915; the construction around 1902 of Ethiopia's first modern roads, much aided by a steam-engine imported in 1904; the founding in 1905 of the country's first bank, the Bank of Abyssinia; the establishment in 1907 of Ethiopia's first modern school, the Menilek II, run by Egyptian Copts; the first modern hospital, the Menilek II, in 1910, and a government printing press, in the following year. The coming of the first motor cars also dates from around that time.

Menilek's modernisation work was continued by his successors, notably after 1916 by Ras Tafari Makonnen, the future Emperor Haile Sellassie. Having obtained Ethiopia's entry into the League of Nations in 1923 he issued the first decree for the gradual abolition of slavery, despatched further students for study abroad, and embarked on a European tour, the first Ethiopian ruler to do so. He later founded the country's first European-style modern school, the Tafari Makonnen, and the first large-scale printing press, the Berhanena Selam, both in 1925, and arranged for the establishment of the country's first airlines and air force.

After assuming the throne in 1930 he introduced Ethiopia's first Parliament, nationalised the Bank of Abyssinia, and founded a national bank, the Bank of Ethiopia. He also established a new national currency, enacted a second and more effective anti-slavery decree, and inaugurated the country's first radio station. The first school for girls was founded by his wife, Empress Manan, after whom it was called.

This modernising era drew to a close in the mid-1930s, when it became apparent that fascist Italy had decided to invade. Thoughts of modernisation, almost inevitably, gave way to thoughts of defence.

247. Construction of the Addis Ababa palace's banqueting hall in the late 19th century. Note wooden scaffolding and temporary stairs made of poles.

248. Ox-drawn carts in early 20th century Addis Ababa market area.

249. Ethiopia's first steam-engine, imported by the Armenian entrepreneur Sarkis Terzian, who had it driven from Jibuti to Addis Ababa.

250. Early 20th century ox-drawn carts crossing the Kassam river, to transport coffee from Sidamo via Ankobar and Harar for export to the coast.

251. Steam-engine imported by Sarkis Terzian, used in Menilek's sawmill for cutting wood.

252. Menilek with one of the first motor cars, presented to him early in the twentieth century.

253. Marking the route for one of Addis Ababa's first roads.

254. Sarkis Terzian's steamroller.

255. Menilek, with a car, in the Addis Ababa market area.

256. The second motor car to arrive in Ethiopia. A German Nacke, belonging to Arnold Holtz, crossing the Kassam river on the way from the coast in 1907-8.

257. The Nacke in the scrublands east of Addis Ababa.

258. A heavy-duty Renault drives down an Addis Ababa street in the 1920s.

259. A group of cars parked outside Addis Ababa's restaurant La Confiance.

260. The railway engine Rhinoceros, manufactured by the Swiss firm of Winterthur.

261. Menilek inspecting the laying of the Addis Ababa-Jibuti railway line.

262. Porters carrying a rail for the railway.

263. Train in front of the Addis Ababa railway station, in construction in 1928.

264. A high steel railway bridge at Shebelé, with a train passing by.

265. Some of Ethiopia's first pilots, with their French trainer Paul Corriger, in 1935.

266. Open air Addis Ababa tailors in the early 1930s, with their Singer sewing machines.

267. Cancelling postage stamps at the Addis Ababa post office, 1935.

268. The Singer Sewing Machine shop in Addis Ababa, 1910.

269. Ethiopia's Foreign Minister, Blatténgéta Heruy Walda Sellasé, speaking on the telephone, 1935.

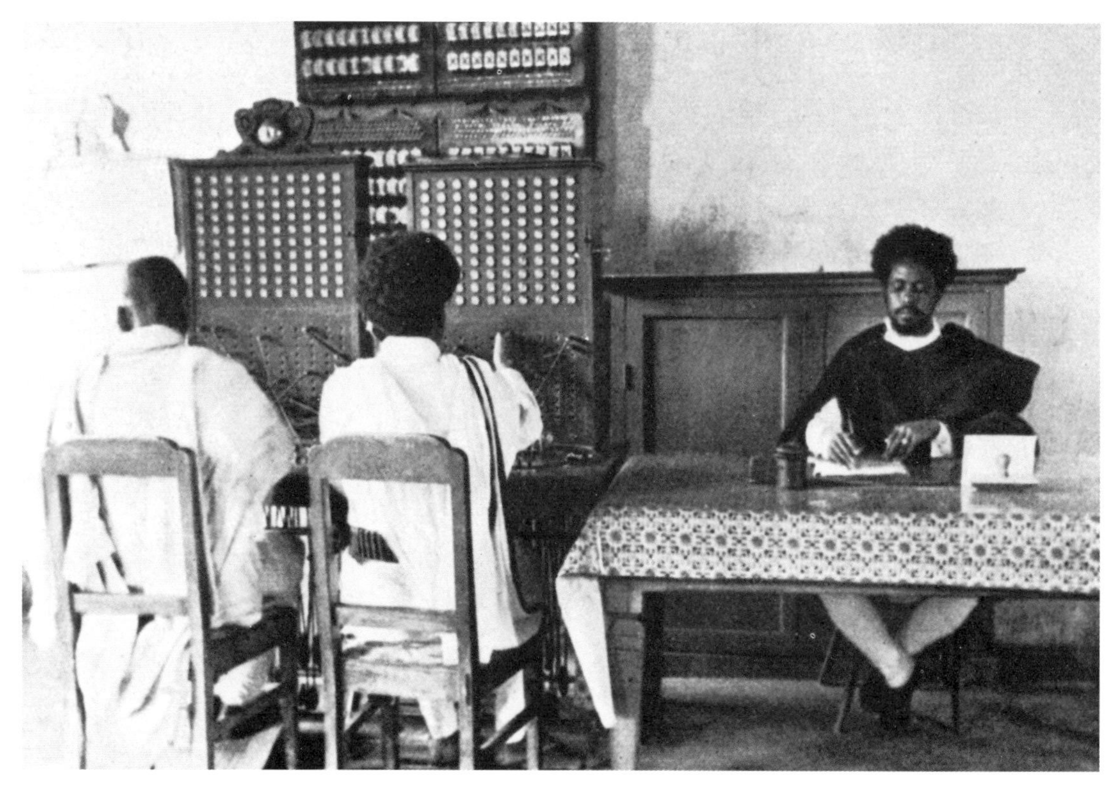

270. The Addis Ababa telephone exchange, with two hundred subscribers in the early 1930s.

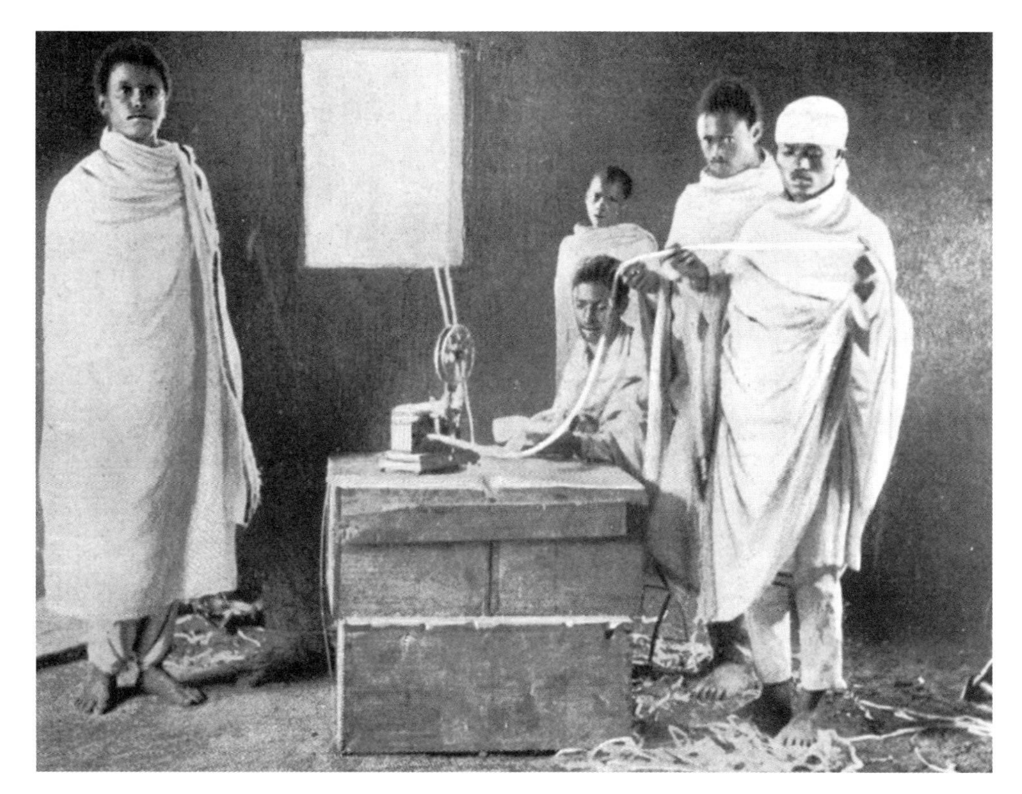

271. Ethiopian telegraphists at work in Menilek's day.

272. Radio station at Aqaqi, 7 kilometres from Addis Ababa, 1932.

273. Menilek inspecting the first Addis Ababa Government school, run by Egyptian Copts. Photo taken in July 1910.

274. Empress Manan visiting Ethiopia's first school for girls, founded by her in 1931, and named after her.

275. The country's first Boy Scouts at drill prior to the Italian fascist invasion.

276. At the blackboard in an early 20th century Addis Ababa school. The student writes in Amharic script, supervised by his teacher, a priest.

277. The Tafari Makonnen School, founded in 1925.

278. Modern vaccination in the early 20th century, replacing traditional variolation.

279. A customer arrives by mule in the early 1930s at a pharmacy run by a German, Hakim Zahn.

280. The Tafari Makonnen Printing Press, established in 1925.

281. Imperial Bodyguard soldiers in front of Emperor Haile Sellassie's newly built palace, 1935.

282. The Imperial Body Guard military band, with its drum-major, 2.8 metres tall, marching through Addis Ababa.

283. An Addis Ababa traffic policeman in the early 1930s. Note placard reading "Vive l'empereur Haile Sellassie".

284. An Addis Ababa policeman. His cap bears the Ethiopian letter "A", for Addis Ababa.

285. Emperor Haile Sellassie addressing the Ethiopian Parliament, founded in 1931.

286. Ethiopian ambassadors, newly appointed in the early 1930s.

VI. Preparing to Resist the Impending Invasion

HISTORIC ETHIOPIA, which had begun to modernise itself at the close of the nineteenth century, was less than two generations later confronted by one of the most serious challenges to its very existence.

The threat had its origin in far-off Italy, where the fascist dictator Mussolini seized power in 1922, and embarked on a policy of militaristic chauvinism. Within ten years of his March on Rome he decided on invading Ethiopia, and despatched one of his closest fascist comrades, De Bono, to the nearby Italian colony of Eritrea to begin preparations for war. Eritrea to the north of Ethiopia and Italian Somalia to the south were rapidly developed as military bases.

Mussolini chose as his pretext for invasion the Wal Wal incident of 5 December 1934. It had its origins in an attempt by the Ethiopian and British Governments to demarcate the frontier between Ethiopia and the then British Somaliland Protectorate. Towards the end of November of that year a joint Anglo-Ethiopian Boundary Commission, operating on the borders of the Ogaden desert, reached the wells of Wal Wal, a hundred miles within Ethiopian territory. They found the area occupied by Italian troops, who, advancing from Italian Somalia, had penetrated far into Ethiopian territory. Italian 'planes that afternoon buzzed the commission, whereupon the British protested at the Italian presence, but promptly withdrew. The Ethiopians, on the other hand, remained, and faced the Italians peacefully, until 5 December, when a shot of indeterminate origin led to an exchange of fire, after which the Italians attacked with 'planes and armoured cars. Mussolini immediately demanded an Ethiopian apology, compensation for Italian losses, and Ethiopian recognition that the Wal Wal was within Italian territory.

The Ethiopian Government responded by requesting arbitration in accordance with an earlier Ethiopian-Italian Treaty of 1928. Italy refused this proposal, whereupon Emperor Haile Sellassie submitted the matter to the League of Nations. The League proposed arbitration, but restricted the issue to the question who had fired the first shots at Wal Wal, which could not be established. The more important question of Wal Wal's sovereignty was excluded from consideration.

Arbitration discussions continued until 3 September 1935, during which time the Italians frantically completed their war preparations. Their invasion took place, without any declaration of war, exactly one month afterwards, on 3 October.

Four days later the League found Italy guilty of aggression, but limited its response, on 18 November, to the imposition of only mild economic sanctions. The list of articles banned to Italy excluded petrol - without which the Italian air force could not have flown. This omission was so significant that Mussolini later observed to his fellow dictator Hitler that if the League had extended sanctions to include petrol he would have been obliged to withdraw from Abyssinia "within a week". The League likewise took no action to close the then British-controlled Suez Canal, through which the Italian troops, and all their military supplies, had to pass.

Mussolini in his propaganda declared his intention of avenging the Italian defeat at Adwa forty years earlier - and of winning for his country "a place in the African sun". There was even talk of peopling Ethiopia with ten million Italian settlers.

The Italians possessed overwhelming military superiority, including unchallenged control of the air. Determined to achieve a rapid victory before the rainy season, due in June or July - and fearing that the League might strengthen sanctions - Mussolini gave orders for the use of mustard gas, which was dropped in bombs or sprayed from the air. This gas, which was banned by international law, did much to break Ethiopian resistance, and enabled the fascist army to enter Addis Ababa on 5 May 1936.

The photographs in this last section exclude any shots of the war - which deserves a separate volume. Our pages are therefore restricted to the last months of Ethiopian independence, a time when Ethiopia tried to re-arm, but lacked the resources to purchase more than a small number of weapons, all of which had to be imported from abroad.

Our last photographs show discussions at the League of Nations; prayers for peace outside St George's Cathedral, Addis Ababa; the Emperor's mobilisation decree; Ethiopian men volunteering to fight the invader, and women preparing bandages for the troops; the Red Cross headquarters in the Ethiopian capital; the departure of soldiers for the war-front - and a solitary Red Cross mule on its way to the unequal war.

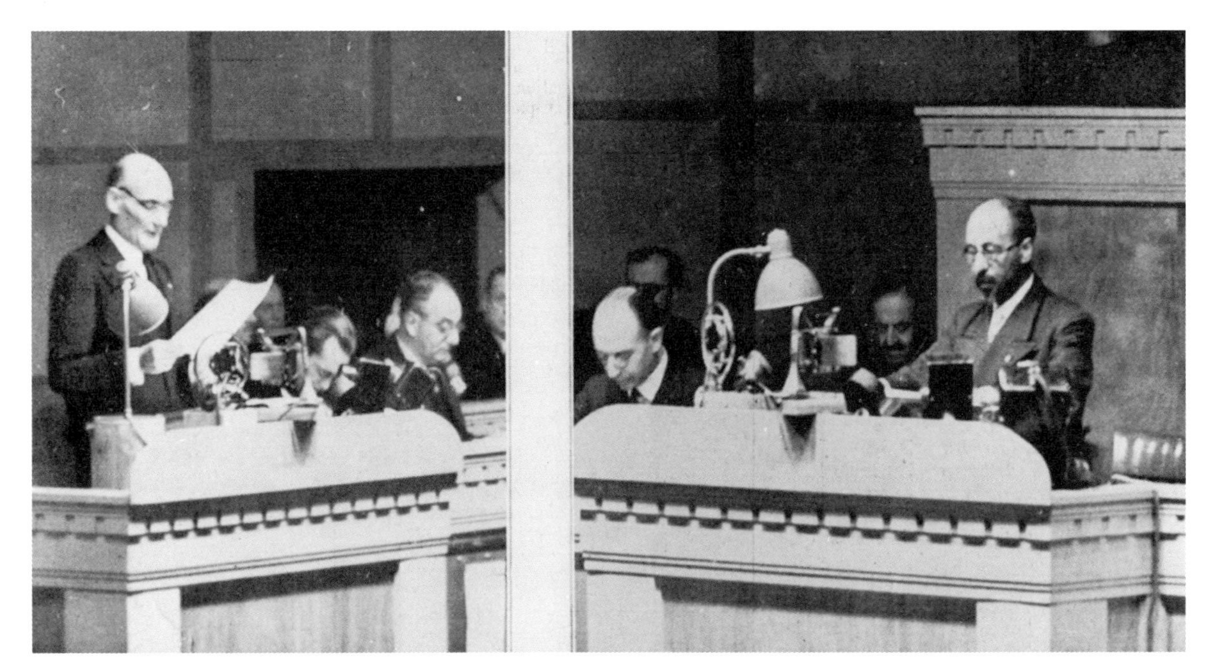

287–8. At the time of the Italian fascist invasion the Italian and Ethiopian representatives address the League of Nations General Assembly in Geneva. Baron Aloisi, left; Bajerond Takla Hawaryat Takla Maryam, right.

289. The Emperor's proclamation of war read out, to the sound of the drum, at Addis Ababa's grand palace.

290. At St George's Cathedral, Addis Ababa, Christians and Muslim's pray together for peace.

291. In front of the Emperor a veteran soldier pledges his determination to fight, and if necessary die, for his country.

292. At the Agar Feqr Mahabar, or Love of Country Society, in Addis Ababa, a recruiting officer calls for volunteers.

293. A soldier at Dire Dawa bids his wife farewell, before leaving by camel for the Ogaden front.

294. The Emperor's Bodyguard seen marching an Addis Ababa street.

295. Ethiopia's poorly armed troops walk through Addis Ababa on the way to war.

296. Off by train to join the southern front.

297. Soldiers and their mules begin their long journey to confront the invader.

298. On the eve of the fighting the Italian minister, Count Luigi Vinci-Gigliucci, the Italian minister to Ethiopia, takes his leave of the Emperor.

299. Troops of Ras Gétachaw Abata camped at Shola Méda on the outskirts of Addis Ababa before proceeding north to Dasé.

300. The tent below the palace at Dasé, from which the Emperor had reviewed his troops.

301. Ethiopia's tiny air force in 1935.

302. Storing grain for the forthcoming war.

303. Importing arms on the eve of the fighting.

304. Three Ethiopian soldiers prepare to resist the invasion.

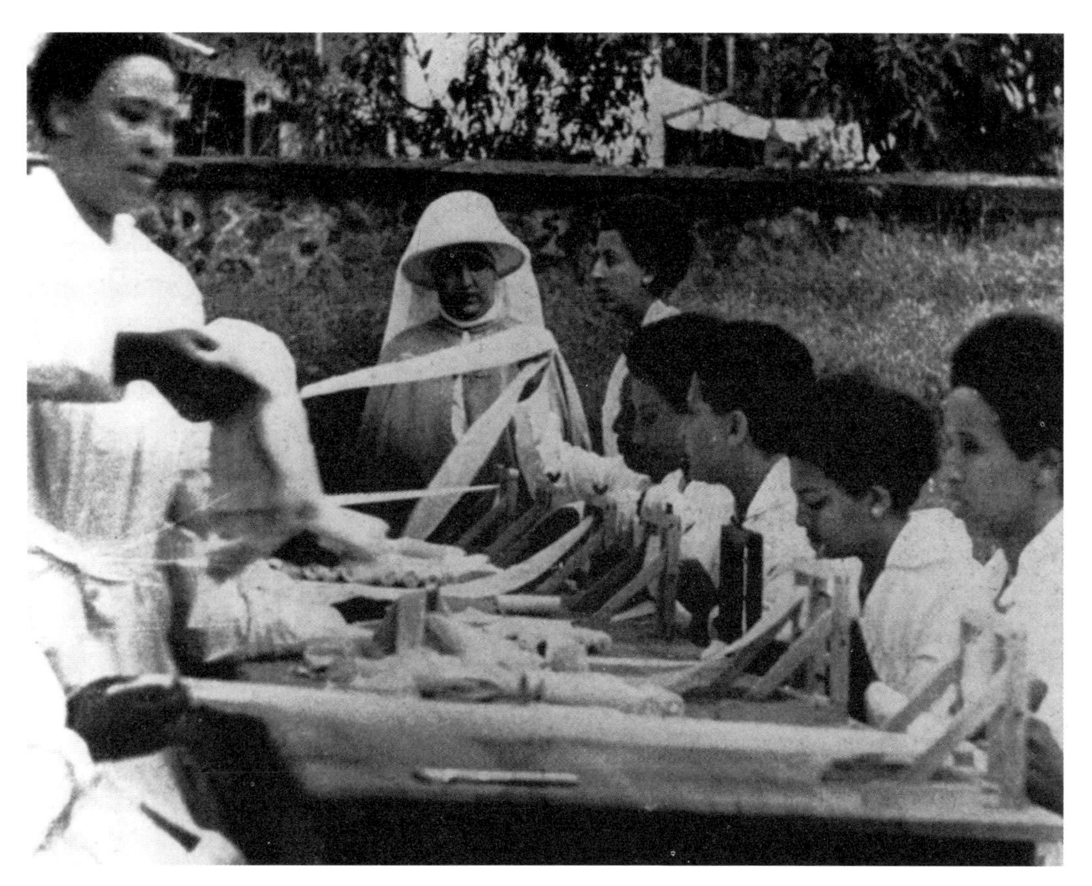

305. At the premises of the newly established Ethiopian Red Cross Society: Empress Manan watches the preparation of bandages. Her daughter, Princess Tsehai, stands at her side.

306. A solitary Red Cross mule goes to war.

PHOTOGRAPHIC SOURCES

Private

Alem Tsehai Iyasu, 204.
Ali Ayo, 105.
Assefa Gabre Mariam Tessema, 59, 63.
Thoros Barssiguian, 14, 45, 66, 77-8, 91, 108, 251, 254, 272, 280.
Boyalu Getatchew, 209.
Desta Solomon, 160.
Patricia Fenn, 35.
Dajazmach Habte Mariam, 104.
Hashim Ahmed Ali, 67.
Alain Le Sear'ch, 83, 85-8, 94, 96, 101, 109, 114-5, 130, 149-50, 166-8, 192, 244, 255, 282, 287-90, 292, 295-8, 300.
Michel Pasteau, 147, 158, 171, 173, 179, 188, 194, 198, 203, 206, 211, 213, 215, 220, 222, 224-5, 229, 234, 240-1, 258, 262-3, 284, 294, 299, 304.
Dajazmach Geneme family, title-page.
Salah ed-Din Mohamed, 68.
Tadele Yidnakachew, 60.
Tekle Tsadik Mekuria, 29, 57, 70.
Wilfred Thesiger, 69.
Yohannes Kinfu, 207.
Many other photographs were lent by private individuals who wished to remain anonymous.

Institutions

Addis Ababa Museum, 268, 273.
British Embassy, Addis Ababa, 5, 197.
" " " , Royal Engineers Magdala Album, 3-4, 16, 18-25.
Institute of Ethiopian Studies, 7, 12-3, 27-8, 33, 38, 42, 44, 51, 56, 117-8, 129, 140-6, 151-2, 154, 159, 161, 163-5, 167, 172, 174-5, 183, 185, 189, 193, 217, 233, 236, 239, 247-50, 255, 261, 264-5, 297.
Royal Anthropological Institute, 62.
Victoria and Albert Institute, 38.

Books

Annaratone, C.A., In Abissinia, 1914, 17, 162, 200, 230, 237, 328.
Azäis, F. and Chambard, R., Cinq années de recherches archéologiques en Ethiopie, 1931, 15, 136-7, 200.
Bruce, J., Travels to Discover the Source of the Nile, 1790, 1.
Cerulli, E., Etiopia occidentale, 1933, 55, 138.
De Castro, L.,Nella terra dei Negus, 1915, 11, 40, 47, 53, 61, 71, 124, 139, 153, 155, 168, 201, 205, 210-2, 235, 245-6, 253, 271, 278.
Coulbeaux, J.-B., Histoire politique et religieuse de l'Abyssinie, 1929, 119, 170, 221.
Duchesne-Fournet, J., Mission en Ethiopie, 1908-9, 133-4.
Emily, J.M., Mission Marchand, 1913, 387.
Farago, L., Abyssinia on the Eve, 1935, 269.
Forbes, L., From Red Sea to Blue Nile, 1925, 76, 81, 99, 113, 177, 243-5.
Goldmann, W., Das ist Abessinien, 1935, 95, 103, 106-7, 132, 198-9, 223, 259-60, 265-7, 270, 275-6, 279, 281, 283, 285, 301-3.
Guèbrè Sellassié, Chronique du règne de Menelik II, 1930-1, 26, 41, 43, 48-50, 74-5, 150.
Hirsch, B. and Perret, R., Ethiopie. Année 30, 1989, 93, 160.
Holtz, A., Im Auto zu Kaiser Menelik, 1908, 30-1, 219, 236, 252, 256-7.
Isaacs, A.A., Biography of the Rev. Henry Stern, 1886, 2.
Jumilhac, Comtesse de, Ethiopie moderne, 1933, 54, 89.
Kabada Tassama, Ya Tarik Mastawasha, 1963, E.C., 98.
Kulmer, F.W. von, Im Reiche Kaiser Meneliks, 1910, 32, 142, 186.
Landor, A.H. Savage, Across Widest Africa, 1907, 190, 208.
Littmann, E., Deutsche Axum-Expedition, 1913, 110-2, 116, 120-3, 127, 229.
Martial de Salviac, Les Galla, 1900, 9.
Martini, F., Nell'Affrica Italiana, 1896, 7.
Montandon, G., Au pays Ghimmira, 1913, 46.
Orléans, Prince, Une visite à l'empereur Ménélick, 1898, 131.
Paulitschke, P., Harar, 1888, 6.
Rey, C.R, In the Country of the Blue Nile, 1925, 99.
Reybaud, L. Chez le roi des rois d'Ethiopie, 1932, 272.
Rosen, F., Eine deutsche Gesandtschaft in Abessnien, 1907, 58, 125, 180-2, 191.
Sa Majesté Hailé Sellassié 1er empereur d'Ethiopie, 1932, 79-80.
Schoenfeld, E.D., Erythräa und äygptische Südan, 1904, 226.
Skinner, R.P., Abyssinia of Today, 1906, 10, 135.
Vivian, H., Abyssinia, 1901, 169, 178.

Periodicals

L'Illustration, 1924, 82-4.
Annales dell'Africa Italiana, 1938, 97.

307. One of 1935 Ethiopia's few anti-aircraft guns.